RAISING ANTI-DOOMERS

How to Bring Up Resilient Kids Through Climate Change and Tumultuous Times

BY ARIELLA COOK-SHONKOFF

balance

New York   Boston

Copyright © 2025 by Ariella Cook-Shonkoff

Illustrations by Pepita Sandwich
Cover Design by Jim Datz
Cover Images © Shutterstock
Cover © 2025 Hachette Book Group, Inc.

Hachette Book Group supports the right to free expression and the value of copyright. The purpose of copyright is to encourage writers and artists to produce the creative works that enrich our culture.

The scanning, uploading, and distribution of this book without permission is a theft of the author's intellectual property. If you would like permission to use material from the book (other than for review purposes), please contact Permissions@hbgusa.com. Thank you for your support of the author's rights.

Balance
Hachette Book Group
1290 Avenue of the Americas
New York, NY 10104
GCP-Balance.com
@GCPBalance

First Edition: August 2025

Balance is an imprint of Grand Central Publishing. The Balance name and logo are registered trademarks of Hachette Book Group, Inc.

The publisher is not responsible for websites (or their content) that are not owned by the publisher.

The Hachette Speakers Bureau provides a wide range of authors for speaking events. To find out more, visit hachettespeakersbureau.com or email HachetteSpeakers@hbgusa.com.

Balance books may be purchased in bulk for business, educational, or promotional use. For information, please contact your local bookseller or email the Hachette Book Group Special Markets Department at Special.Markets@hbgusa.com.

Print book interior design by Amnet ContentSource

Library of Congress Cataloging-in-Publication Data

Name: Cook-Shonkoff, Ariella, author.
Title: Raising anti-doomers: how to bring up resilient kids through climate change and tumultuous times / by Ariella Cook-Shonkoff.
Description: First edition. | New York, NY: Balance, 2025. | Includes bibliographical references and index.
Identifiers: LCCN 2025017072 | ISBN 9780306833571 (hardcover) | ISBN 9780306833588 (trade paperback) | ISBN 9780306833595 (ebook)
Subjects: LCSH: Parenting. | Anxiety in children. | Social problems—Psychological aspects. | Climatic changes—Psychological aspects.
Classification: LCC HQ755.8 .C658 2025 | DDC 649/.1—dc23/eng/20250528

LC record available at https://lccn.loc.gov/2025017072

ISBNs: 978-0-306-83357-1 (hardcover); 978-0-306-83359-5 (ebook)

Printed in Canada

MRQ

Printing 1, 2025

For my daughters and the next seven generations.

"For there is always light if only we are brave enough to see it. If only we are brave enough to be it."

—Amanda Gorman, American poet and activist

Contents

INTRODUCTION
An Unabashed Rallying Cry for Parents Everywhere
1

PART 1: PARENTS

CHAPTER ONE
Parenthood: A Major Life Reboot
15

CHAPTER TWO
How Anxiety Operates in Families
41

CHAPTER THREE
Creating a Calm and Healthy Home
61

CHAPTER FOUR
Why Parents Bury Their Heads in the Sand—and How We Can Look Up
91

PART 2: KIDS

CHAPTER FIVE
Emotions "R" Us
125

CHAPTER SIX
A Twenty-First Century Anti-Doomer Toolkit for Families
147

Contents

CHAPTER SEVEN
How to Talk to Your Kids Amid a Brewing Existential Sh*tstorm
205

CHAPTER EIGHT
Growing Your Identity: Practices for Staying Engaged
237

PART 3: COMMUNITY

CHAPTER NINE
From Alpha to Boomer to Z: Strengthening Intergenerational Work as a Way Forward
259

CHAPTER TEN
A New Vision of the Future: What Happens When We Sweat the Big Stuff, Do Right by Our Kids, and Stay Committed
291

Appendix A: Sample Automatic Negative Thought (ANT) Log   *307*
Appendix B: Self-Care Weekly Tracker   *309*
Appendix C: Tools for Managing Anxious or Traumatized Nervous Systems   *311*
Appendix D: Indigenous-Focused Resources   *313*
Appendix E: Resources for Connecting Low-Income and BIPOC Communities with the MHW   *317*
Appendix F: Climate and Mental Health Resources   *319*

Acknowledgments   *323*
Notes   *329*
Index   *337*

Introduction

An Unabashed Rallying Cry for Parents Everywhere

*I*n 2014, I stood waiting in an airport security line with my partner en route to our honeymoon in Kauai. A young couple behind us smiled, which I attributed to my visible baby bump.

Before my manual pat down, the man behind us said: "Congratulations! Y'all are about to enter one of the biggest, most awesome clubs in the world—the Parent Club."

My partner and I looked at each other, then back at them, as I eked out a polite giggle.

"Thank you," I said, appreciating what I felt was a kind remark, even though I didn't really know what it meant.

We exchanged a few more pleasantries, but those words about the "Parent Club"—welcoming, yet mysterious—stuck with me like glue.

Crossing that tender threshold from pregnancy to new motherhood, and watching my first baby and then the next one grow from infancy to toddlerhood to school age, the significance of those words held fast. Even as I struggled at times along the way, I felt relieved that I was not alone. I was part of a club, a circle, a crew; a cooperative fellowship, a coalition united in purpose: doing the best for our children.

As I navigated the emotionally textured first years, I found myself hungry for conversation and connection with other parents. A magnetic attraction pulled me toward them, an instant kinship sparking even during the most mundane conversations: *Is she sleeping through the night yet? Does she take a pacifier? Watch out, she dropped her lovey!*

Gradually, I began to discern a certain twinkle in a parent's eye that says without any words: *I'm a parent, and I've been through the "trenches," and we might have some similar experiences, so you can probably relate to me, and I to you, to some degree.*

Parenthood is a humbling, bring-you-to-your-knees type of life transition, and cuts across geography, culture, ethnicity, class, and sexual orientation; it's honest, binding, and transcendental. And let's not forget one more vital and unifying thread: parents generally want to do good by their children. The inherent vulnerability in parenthood creates unexpected opportunities for broadening perspectives, forging new connections, and a sense of solidarity.

It's a good thing that there's a Parent Club out there, since we as parents have a difficult world-maze to navigate. The environment we are bringing our kids into is not the same one that we were born into. We have to help them—and ourselves—figure out not just how to grow up, but also how to grow up in a world that can be violent, polarized, oppressive, and degraded.

While the initial adjustment to mom life was not particularly easy for me, I was fortunate to go through the transition before the additional worries of COVID-19 arrived. Motherhood, for me, entailed blinders, and my world shrunk in size to the essentials: our kitchen, our bedroom, our bathroom, our baby's room, and our neighborhood.

My struggles, which felt so insular at the time, despite their universality, consumed me. *Would I ever sleep for eight hours straight again? How*

Introduction

come establishing a life balance felt so damn unattainable? Would my partner and I see things the same way when it came to parenting?

Then one morning, when my oldest daughter was just shy of two years old, fluttering ash and thick smoke arrived on our doorstep. Overnight, it was no longer safe to take my daughter out for a walk in the stroller, to let her swing at the local playground, or to drink tea on our patio. The climate crisis—which to me at the time seemed always to be assigned to some far-off date in the future by politicians, teachers, or scientists—was here, knocking at our door.

Parenthood, by definition, is laced with all kinds of anxiety. We worry about the developing fetus during pregnancy. We worry about labor progressing smoothly. We worry when our kids are sick. We worry if they're struggling with school or in their relationships with peers. We worry if they're not home by curfew. We worry when we're alone in the empty nest. We worry, we worry, and we worry.

However, the day that wildfires arrived, my eyes opened to a visceral, felt sense of climate change and it ratcheted up my sense of vulnerability as never before. A couple of years passed, followed by the birth of our second daughter. Suddenly, there was something called "wildfire season" that descended each year. Suddenly, there were test evacuation calls coming through in case of a wildfire—not just smoke—coming to our own neighborhood.

When the electric utility company shut off our power for a few days as a precautionary measure, I packed my kids' lunches and sent them off to school. At night, our family cooked dinner out of a giant ice cooler and crouched around the coffee table in the family room playing card games by candlelight. My eco-anxiety thrummed a low-grade apocalyptic drumbeat. Bum-ba-dum-ba dum ba dum.

As a therapist, I know that this cognitive dissonance—acting and behaving in opposition to your thoughts and feelings—feels robot-like and bizarre. You see red flag warning signs around town, your body bristles in the dry, gusty winds, but you pack unicorn and Hello Kitty lunchboxes and drop off children at school anyway. And repeat. Leaving town to attend a workshop under these circumstances was difficult. But time and space of my own—a parent's antidote to parenting—was a gift that ultimately led me to greater clarity.

Driving north to Oregon, I passed charred forests and remnants of controlled burns while listening to news about wildfire evacuations in Southern California. My heart sunk as I slowed to take in the acres of trunks blackened by the notorious Camp Fire in 2018, which caused eighty-five deaths and destroyed more than eighteen thousand structures. Smoke-tinged air blew through my windows. Gone was the usual adrenaline thrill I felt from a road trip. That low-grade anxiety remained in the pit of my stomach every mile to my destination.

Then I started recording my thoughts and feelings on tape.

A few days later, on my return trip, I talked aloud and recorded some more. I knew what I was feeling was called "eco-anxiety" and that, while it was raw, it was also a perfectly healthy and rational response to the changing environment outside of my window. And if I was feeling it, I suspected that many other people, especially parents, were feeling it too. This anxiety—this worry and fear about the besieged world our children will grow up in and one day inherit—must be a common experience among millions of members of the Parent Club. As I drove toward a red-violet sunset horizon, in my mind I'd already begun writing my story.

Sitting in my office with clients was a different experience. When nearby wildfires raged, smoky air penetrated the window glaze while an air filter blasted, and uneasiness stirred in the pit of my belly. I didn't feel safe. My kids were in the back of my mind, and I worried about the possibility of a wildfire sparking in my own community. Yet I was determined to transition smoothly back into work after my second maternity leave, so I compartmentalized my climate anxieties as best I could.

My clients came and went, and hardly anyone mentioned wildfires or climate change. If I'm being honest, I felt both a sense of relief and

Introduction

frustration. Relief in the sense that I felt vulnerable in tackling this daunting subject and wasn't even sure what to say if my clients brought up the elephant in the room. Frustration in the sense that I knew continued avoidance of climate change was dangerous—potentially even lethal—and signaled that deep psychological defense mechanisms might be at play. My adult clients in particular (not unlike me!) clung to habitual avoidance, denial, and disavowal, nary a one mentioning the ecological crisis unfolding outside the office window, stinking up the room. Who was I to judge?

Early in our careers, therapists learn to skillfully manage our responses to clients. This helps to ensure a safe, calm, and nonjudgmental space. Yet what happens when climate change threats multiply, pose an existential threat, and remain deeply mired in the unconscious? There was certainly no playbook for this in grad school. To ignore or brush aside the climate emergency and fail to invite space for processing climate emotions in sessions could be viewed as colluding with our clients in their defenses.

I found comfort in connecting with a growing number of mental health professionals, climate-aware therapists, who help clients acknowledge and explore the elephant in the room. As psychologist Elizabeth Allured puts it, climate-aware therapists "hold in mind the larger environmental crisis we're in, allowing it to surface into consciousness."[1] There's an ethical impetus for therapists to support clients in turning toward difficult truths, in moving past "stuckness" or other defenses that are not serving them. So why would we not do the same with the climate emergency? And when we look at the climate emergency, boy, does that qualify as an existential threat.

As a mom, I knew we needed something else too. We need a parent fight club. This book is a handbook for parents wanting to join this parent fight club, for those parents who no longer want to turn a blind eye to reality but instead show up for their kids in concrete ways. By embracing the reality of climate change and other stressors, we can harness the true potential of the Parent Club to ensure a healthier, safer, greener, and more equitable future for our kids.

In writing this book, I hope to address the looming and often unnamed emotional burden that many parents are carrying alone right now. As a good friend of mine put it: "I walk around with a profound fear lurking inside of me: Are my children f*cked?"

Many of you are grappling with this question in your own way, whether you're aware of it or not. And other questions mushroom out, like: Is there a climate resilient place we should move to? Should I equip my kids with survival skills for societal collapse? Will my kids live a life of anxiety and fear? Will they have a future that's worth living?

One parent told me: "The fear that I carry is so huge—I wall it off and hold it in, and then get anxious about smaller things. What I'm actually scared of is what my kid's life is gonna look like? . . . I don't like living like this."

It's true that parents have encountered existential threats and crises—and endured the emotional fallout—over the centuries, and that parenting has always been a rigorous, eat you up and spit you out, unpredictable emotional endeavor. Yet at the dawn of the twenty-first century, there are indisputable challenges that are worth examining for their particularities, and we shouldn't gloss over or minimize them. I'm talking about the confluence of fascist or fascist-leaning governments (and the resurgence of fascism overall); the powerful propaganda and disinformation campaigns easily disseminated over social media; the weakening of democracy in the US by a former jilted president (who won a second term as a convicted felon); the unleashing of AI despite ethical considerations; the proliferation of screens and technologies that distract and distort human lives; the approach to and surpassing of multiple climate tipping points; the global rise of antisemitism and neo-Nazism; a violent and volatile Middle East; and the list goes on.

The more thoughtful reframe here is that parenting has been and always will be a difficult enterprise across history, and each century brings

Introduction

its distinct challenges. Today, many of these challenges are abstract and existential, adding a new layer of complexity and confusion in how to raise kids, one that is quite different from running from a tiger with a baby in your arms in the wild or even sending a child to school where they are bullied. With threats that are covert (social media sites, marketing algorithms), invisible (rising atmospheric carbon dioxide levels), and unpredictable (mass shootings), children are exposed to a multitude of manufactured, adult-induced harms.

These are some of the intense, gripping fears that many parents live with today. I'm about to introduce stuff that you might already know or might make your stomach drop, but hang in there. If we don't face reality, then we certainly can't change it. And if we don't stop to think and learn about reality, then our kids will know more than us and our ignorance will be their burden.

I'm not saying this will be easy or that you won't encounter potholes or roadblocks along the way. But I do promise that in reading this book you'll gain insight, reorient to today's challenging socio-political landscape, sharpen your tools for staying balanced and resilient, craft a parent identity that will keep you on track, and feel comfortable to broach tough topics with your kids. Out of adversity comes strength, and the opportunity to become even more awesome parents. We need our togetherness more than ever right now to contend with the omnipotent forces that be.

Before I go on, I must acknowledge the honest, nagging question that's on many parents' minds right now regarding climate change: *So, how bad is it?* Here are three concrete statistics to help you see more clearly. You can revisit these or update them periodically to help maintain your edge and energy.

1. The top ten hottest years ever recorded in Earth's history have all been after the year 2000. The annual global average temperature is 1.1 degrees Celsius hotter than preindustrial levels and could reach

1.5 degrees Celsius as soon as 2040 or earlier, at which point climate impacts are predicted to severely worsen and be irreversible.
2. Several climate tipping points—thresholds that when exceeded can lead to irreversible change—may be crossed this century due to human activity: ice sheet collapse, destabilization of the Amazon rainforest, and compromised ocean circulation are some examples.
3. Per the Intergovernmental Panel on Climate Change (IPCC) 6th Assessment Report, around 3.3 to 3.6 billion people already live in contexts that are highly vulnerable and impacted by climate change. The report also says that 50 to 75 percent of the world population can be subjected to periods of "life-threatening climatic conditions" by 2100. Vulnerable groups most at risk include women, children, the elderly, ethnic and religious minorities, Indigenous people, refugees, and low-income and agricultural communities.[2]

In short, it's not good. The Parent Club should be worried. But to quote sustainability scientist Dr. Kimberly Nicholas: "It's warming, it's us, we're sure, it's bad, we can fix it."[3]

I want you to feel comfortable and supported as you embark on the journey ahead, so here are some additional notes to prepare you.

- The book is divided into three parts: "Parents," "Children," and "Community." You'll notice that much time is spent on what is sometimes called "internal work," or emotional processing, because going inward is ultimately your way to negotiate strength, balance, and calm, as ironically as that may sound. I know that some of you might think, *Well, that sounds great, if I had the time.* I'm here to say that this isn't just some extra credit privileged shit but rather that it's essential for sustained long-term engagement. And it's work that you'll come back to over and over again. By the end of the book, I'll have laid out three paths for you: Will you be a doomer? A hopium

Introduction

addict? Or an anti-doomer? And, more importantly, what will this mean for your kids?
- Regarding language choice and inclusivity: To communicate and delineate particular parental experiences throughout this book, I use gendered terms such as "mother" and "father." However, it's important to recognize the limitations of these terms, which do not reflect everybody's pregnancy or child-rearing experiences, particularly those whom identify as nonbinary or transgender, and who must navigate often exclusive, heteronormative language within the medical and political systems.
- The Parent Club includes any parent, caregiver, guardian, foster parent, involved family, or community member. Although I frequently reference American culture, this book extends to the global village of parents. After all, what we need right now is a global Parent Club coming together in community, in solidarity, and in strength.
- Throughout the book, I encourage checking in with an "accountability partner" or "Climate Buddy," a person you trust, can lean on for support, commiserate with, and be vulnerable with. When you have a partner, companion, or group, it helps you to stay engaged. When one of you falls off the climate wagon, the other can be there to help. If possible, try to identify someone before you dig in. (And please be mindful of appropriate professional or other boundaries.)
- I use a sprinkling of acronyms throughout, and it's worth explaining MHW, which stands for "more-than-human world." This phrase, coined by eco-philosopher David Abrams, refers to an all-encompassing realm of earthly life that includes but goes beyond human culture (so as to embed humans more firmly in the natural world).
- In this book we'll practice cultural *appreciation* by identifying traditional sources of knowledge, and by relating to and finding inspiration in practices different from our own. Please be mindful about overstepping or understepping into cultural appropriation. I'll say more about this in chapter 6.

- At the end of each chapter, you'll find an art therapy prompt. Even if you don't consider yourself an artist or artsy, or doubt that you have the time, I still encourage you to give it a try—it could truly be transformational for you, as it is for many of my clients. Yes, it might feel vulnerable, scary, uncomfortable, or hokey. Try thinking of art-making as a tool for self-exploration, where there's no actual right or wrong. By giving yourself permission to create, you'll move past defense mechanisms to tap into your unconscious. What bubbles up inside of you—thoughts, sensations, and emotions—is often as revealing, if not more, than the art that you produce. Sometimes a low-stakes warm-up is helpful, like scribbling on paper with both hands or shaking your sillies out to music.
- I would encourage you to consider setting up a mini art studio. This could be in a corner of your home or office, on a bench or table outside, or in a garden or park—anywhere that you have some degree of peace, quiet, and safety. This might be an easy task or it might be a huge feat, depending on your circumstances. Try gathering various supplies that you may have on hand, repurpose or reuse, or find at local thrift shops, salvage yards, garage sales, or art stores: different size and textured papers, cardboard, scraps of wood, canvas, clay, glue, scissors, scraps of yarn, tissue paper, old magazines. Try to have both wet (e.g., paint) and dry (e.g., markers, oil pastels, colored pencils) media available. For more tips, please visit my website: www.ariellacookshonkoff.com. Do not be scared off by this step; the set up can be as basic as you'd like!
- Sprinkled throughout this book are Parent Pauses and Parent Centering Practices (PCPs)—moments when you'll move out of left-brain didactic learning to right-brain experiential doing. These reflective exercises are an integral part of this book. These exercises are not to be rushed or ignored, so try to reserve time for them or come back to them later.
- You may be wondering, *Can I read this book without participating in the Parent Pause or the art directives?* Well, sure, absolutely. However, you'll miss the chance for deeper emotional transformation, and

Introduction

you'll bypass some of the avenues for resilience offered in this book. I encourage you to try some of these exercises even if they are not your usual jam.
- I'll remind you periodically to self-regulate at intervals because this material can feel heavy at times. But please don't wait for permission—honor your instincts. You can set the intention or mood for a PCP by doing something calming like lighting a candle, wrapping yourself in a blanket, playing music, brewing tea, or moving outside into nature.

Addressing tough real-world issues while staying engaged as a parent, and even reading this book cover to cover, is not particularly convenient. I get it. A million other things are screaming for your attention. To add "do something about the climate emergency" to your already unwieldy list can feel like more pressure than necessary.

But to reiterate: Climate change is not theoretical but a ginormous global in-your-face emergency that requires immediate action. Based on the findings of the IPCC 2021 report, the UN Secretary General issued a "code red for humanity."[4] Following the IPCC 2022 report, the cochair of the working group announced that it's "now or never."[5] Following the the IPCC's 2023 report, the UN Secretary declared "the climate time-bomb is ticking."[6] Each year, scientists and experts are screaming for our attention, and trying to find the words to get the message to sink in. In parent terms, this translates into "Dial 911" or "1-2-3 . . . All eyes on me!" We desperately need cooperation, and we need a crisis response.

Beyond our personal crises, there's no more urgent concern right now than the climate crisis. And I say that with the caveat that the climate crisis is part of the polycrisis—a confluence of overlapping existential stressors—and it's not pigeonholed to halting carbon emissions in the atmosphere. Addressing the climate emergency is also about rectifying the egregious inequities baked in. We can do this by increasing representation of Black, Indigenous, and People of Color (BIPOC) communities on the frontlines of decision-making, by combating environmental racism in our communities, and by making reparations to countries who have contributed the least to global warming but bear the harshest impacts of it. Let's not forget the countless faces and names of kids, parents, and community members who are caught up in this human-caused mess all around the world, even as we speak in such generalities. Let's also not forget about those who have perished because of the climate crisis—or will perish in the years to come.

As a cisgendered, white, heterosexual, Jewish American embedded in American consumerism, having the time and space to even *think* about climate anxiety is a privilege that I own. But more than that, it's a privilege that I know I must act on.

I'd like to put forth the notion that those of us with some degree of privilege share a responsibility in doing what we can to enact systemic political change. Not every parent is free to exercise their voice, particularly those who are oppressed or marginalized. When we have opportunities to speak up, act, partner with, or amplify marginalized voices we must do so.

Although the change we need to see is systemic and global—not just personal—we start by planting seeds wherever we are: in our homes, schools, and communities. Then, do you know what we do? We go BIG. The rumor out there is that many people are scared of and prefer not to deal with angry parents. So let's do it. And let it be the fight of our lifetimes, for the sake of other lives and lifetimes.

If that sounds like a rallying cry, it is, but remember, just as we tell our kids: small steps first. My hope is that this book will resonate with an emotional chord inside of you, stirring you awake or into action, and offering you a lifeline of resources and ideas.

Whether you're a parent, teacher or engineer, an artist or scientist, in IT or HR, sales or in a union, a stay-at-home parent or currently unemployed, and whatever your religious or political affiliation (or lack thereof), there's not a moment to lose. This book is for you, and the Parent Club is waiting.

PART 1

Parents

ONE

Parenthood

A Major Life Reboot

"I love my child, but the first year was hell."
—Parent Club member

In an instant your life changes forever. One day you're pondering or speculating about what it might be like, and then suddenly it's day one of parenthood. You can feel that remarkable shift inside of your very bones.

Once baby arrives, there ain't no stopping the clock; life unfolds around you while you try to sort things out on the fly. You'll never regain that sleep. You'll have to find the time to get that house project done during your baby's naps. You might find yourself "sandwiched" between caregiving for aging parents or ill family members and children at the same time. There's no restart. Whether you're taking care of an ill family member, coping with an unexpected layoff, starting a new job, separating from a partner . . . there you are, with your child to also consider. It's full on. Welcome to parenthood!

One lesson I received early on: just as you start to understand a particular stage that your kid is in and gain some semblance of equanimity, it will change. The rapid changes that kids go through are awesome to behold, but they can also be stressful and vexing. Once you finally feel like you're getting your bearings, the ground beneath you shifts. The ancient Greek philosophers didn't exactly spend time pondering about the good life as a parent. But we can borrow a basic truth from Socrates: "The only thing I know is that I know nothing." Indeed. What a mantra for life as a parent. We might as well remind ourselves of this every year on our child's birthday.

Many parents are eager to hightail it through those rocky moments when a child's behavior eclipses all sense of household normalcy. They are desperate for support, stability, and reassurance. Part of my job as a therapist is reminding them, ever so gently, that this is normal, that this is all in their job description, and will be for a long, long time. Of course, my job is also to reassure them that they can do this, to help them build out their toolkits, perhaps open their minds to seeing things in a fresh way, and resource them to resume effective leadership in their families.

The difficulty here is that our kids aren't the only ones changing. We are too. And that realization starts during pregnancy and blows up somewhere in that period of post-childbirth adjustment.

An Incomprehensible Milestone

When you begin operating with a plus one in mind, life choices and decisions feel different. Your old value systems, habits, and behaviors start to shake, tremble, even collapse. Your ego gradually starts to make its way to the backseat, relegated by your kiddo's best interests.

Suddenly, it's time to assess trade-offs and quality of life. How will I make enough money to support my family? Maybe it's time to live closer to the grandparents? Do I really want to commute for X hours each week? Should we leave the US and raise our child where school shootings aren't as common? Should I cancel my social plans because family dinners are important? . . . The list goes on. (And, of course, these are privileged conversations to have for sure.)

When I was in my twenties, I rode public transit in San Francisco for hours each day. I lived near the beach and had to take several forms of transit to get to my design classes on the opposite side of town. If I wanted to have a social life, I'd commute all over again in the evening. I did this for years: walking and biking were part of my lifestyle, and it was a great way to get to know the city.

As a parent, I've had to up my efficiency game. With a crying baby strapped into the backseat, or two squabbling sisters, I have to admit that getting from Point A to Point B in a timely fashion, which increases the chances of prolonged harmony, is a current value. Whereas I used to meander around the city in a dreamily spontaneous way, these days I'm a more streamlined and goal-oriented version of myself.

Other examples of shifting values in parenthood include aspiring to be a good role model, giving up an addiction or bad habit, working toward work-life balance, raising kids in a diverse community, and teaching them how to live an ecologically sustainable lifestyle. Even as we grow into our roles as parents, we might not recognize our moods, behaviors, the words that come out of our mouths, how we show up in relationships, or even what our priorities are. For example, we might never have predicted our future life as a room parent or soccer coach, guessed that we'd adopt a "helicopter" parenting style, that we'd allow certain friendships or hobbies to fizzle out, or that we'd adopt our own parents' behaviors. Indeed, the parenting enterprise is unpredictable and we are not in total control, no matter what we might tell ourselves.

Another fundamental shift around midlife is a desire to help guide the next generation, to leave a positive imprint on the world. This reckoning with meaning, purpose, and morality is what developmental psychologist Erik Erikson called "generativity." While there are many ways to make a lasting contribution, having a flesh and bone child is a concrete legacy. You might feel a natural inclination to get involved in some way, such as volunteer at your kid's school, help them stand up to bullying or racism, or teach them a skill, like fishing or cooking.

Perhaps adjusting to parenthood would be easier if this major life transition got the airtime that it deserves. There are some ways that we do this, and some ways in which it's sugarcoated or ignored. While there's a certain commercialization around motherhood with its emphasis on baby showers and gifts, there has been far less exploration of a mom or dad's emotional experience.

This is unfortunate. Relying on Hollywood movies for answers does not do us any favors. We need honest real-life examples.

The parenthood milestone begs to become more ritualized and acknowledged in American culture—much like a wedding, quinceañera, or bar/bat mitzvah. The resurgence of anthropological terms *matrescence* (transition into motherhood) and *patrescence* (transition into fatherhood) certainly aid this conversation. But we need to go further to embed these into cultural relevance. Marking the Parenthood Milestone lists some of the rites of passage or traditions to celebrate the initiation into the Parent Club. What are some ways you might have done this? Some ways you might do this in the future? Some ways you might help others to do this?

MARKING THE PARENTHOOD MILESTONE

- **Baby shower:** This can include rituals that not only celebrate *baby's* arrival, but also to acknowledge the making of a new mother or father. You might try belly painting, stringing a necklace or bracelet with beads from each guest to wear during childbirth, a belly plaster cast, or a memory book for guests to sign.
- **New mother or father's circle:** Each guest offers reflections, insights, hopes, and dreams for a new mom or dad; a form of witnessing and initiating new parent and establishing community.
- **Symbolic rituals:** Herbal bath, postpartum ceremony, massage, placenta burial, and tinctures.
- **Blessingway ceremony:** A Navajo traditional rite of passage during pregnancy that focuses on spiritual and emotional support.
- **Godh Bharai:** A traditional Hindu practice in which expecting moms are adorned with henna on their belly, hands, and feet.
- **Closing the Bones ceremony:** A South American post-childbirth ritual that includes massage and body wrapping (or *rebozo*) to help stabilize the pelvis and hips and support spiritual recovery.
- **Sitting the month (*zuo yue zi* in Mandarin):** A traditional Chinese practice of postpartum recovery in which new mothers stay indoors for a month to heal and bond with baby.
- **Mikveh:** A spiritual bathing ritual in the Jewish tradition after childbirth.
- **Babymoon:** An intentional bonding trip, retreat, or vacation for expecting parents. While some babymoons are lavish vacations, this can be as simple as carving out an intentional block of time to connect with a partner or friend about hopes and fears.

Indeed, marking this milestone can be a helpful way to slow down and acknowledge your entry into the Parent Club. But it's also important to acknowledge the harder parts of becoming a parent. Indeed, the transition into parenthood is hardly one-sided bliss and joy; there are plenty of hardships and pain points as well.

What's Underneath All of the Hearts and Rainbows? Trauma and Grief

If we don't do our own work to heal ourselves, then we are destined to pass along our own traumas to our kids. This happens all of the time; in the mental health field, we call it intergenerational, or cross-generational, trauma. In fact, severe trauma may even be passed on through genes. Children of Holocaust survivors—and children who were in utero during the time of the World Trade Center attacks—both revealed lower cortisol levels.[1] (While lower cortisol might seem like a good thing, because it is a stress hormone, it can lead to negative adrenal health issues.) To get on the same page, let's define trauma as "a lasting emotional response to a distressing life experience."

Think about how a mom recovering from a traumatic birth would feel differently than a mom who experienced a self-reported "positive" birth experience. Not only will this affect her postpartum mental health, emotional adjustment as a new mom, identity and possibly confidence as a parent, but it also may impact her physical recovery, her experience with nursing, maternal bonding, and so on. Or, think of a parent who is abandoned by their partner while pregnant or soon after childbirth and will now have to raise the baby alone. Or, consider a mother who endures a difficult or anxious pregnancy after having previously experienced a late miscarriage or stillbirth. There are countless examples of trauma, many of which receive little emotional acknowledgment, let alone time to heal. But to help kids deal with all the trauma in the world today, we must start attending to our own healing as well.

With that in mind, let's take a sober look at trauma and grief within the experience of parenthood. The term "perinatal" typically denotes the

Parenthood

window of time a few weeks before and after childbirth—a particularly vulnerable one for moms-to-be. Postpartum Support International—a nonprofit that promotes mental health awareness related to pregnancy and postpartum—extends this term to encompass all of pregnancy through the first year of a baby's life, roughly a period of twenty-two months. This is a much-needed redefinition! Not only are the major changes in pregnancy recognized, but the postpartum shifts are as well.

What a lot of new parents miss amid the ooh-ahhing, new baby smells, and sleepless nights is getting in touch with their grief. Even if you don't think that applies to you, it's an inherent part of new parenthood. What was gained? What was lost? How were your expectations met or unmet? (In case you haven't carved out time to reflect on these questions, not to worry, we will do so at the end of this chapter.)

Now there is a grief and then there is Grief, with a capital G. When birth outcomes are not as expected, and we find ourselves blindsided, underprepared, shocked, confused, challenged, or flooded, we will eventually face grief one way or another. As a culture, we are quick to overlook or minimize traumas that may have occurred during pregnancy, childbirth, or postpartum. When baby arrives, they are catapulted onto center stage. Mom is often pushed to the periphery as a supporting actor.

Years ago, a teenage client of mine arrived at our program office with her new baby and was immediately surrounded by a flock of crowing peers and therapists, all clamoring on about how beautiful her baby was. I recall her looking rather shell-shocked, with heavy bags under her eyes, and a polite smile on her face.

Later she told my colleague: "Everyone always asks me about my baby. No one ever asks me how *I* am doing."

Of course, grief and trauma are not necessarily meant to be processed in hospital recovery rooms. These matters deserve time and space. Mama's groups, parent groups, perinatal mood and anxiety groups, individual therapy, body work, retreats, sharing with those you trust, time in nature—are all of these are supportive options.

Capital "G" grief goes hand in hand with capital "T" trauma; wherever there is significant trauma, there is also significant grief to process. But there are plenty of lower case or more subtle examples that also affect parent minds, emotions, and bodies. (See Examples of Birth-Related Trauma/Grief.)

EXAMPLES OF BIRTH-RELATED TRAUMA/GRIEF

- **Pre-term birth**
- **Stillborn birth**
- **Miscarriage**
- **Abortion**
- **Fertility issues**
- **Infant-parent separation**
- **NICU stay**
- **Unexpected maternal medical complications**
- **Unexpected infant medical interventions or health conditions**
- **Deviation from a birth plan**
- **Pain medications administered, unexpected or undesired**
- **Delayed bonding time with mom or dad, or separation for prolonged period**
- **Mother who did not feel adequately supported or advocated for during labor and delivery**
- **Mother who was unable to communicate her birth preferences, or did not feel heard or respected**
- **Breastfeeding difficulties**
- **Unwanted baby**
- **Baby borne out of sexual abuse**
- **Domestic violence or substance abuse during pregnancy**
- **Child Protection Services involvement during pregnancy or after birth**

Parenthood

Although traumatic birth experiences are common, we often don't hear those stories because mothers feel afraid to speak up due to shame or judgment, or they simply don't want to think or talk about it. A client of mine voiced her disappointment in how labor and delivery progressed as she experienced intense pain, fear, and physical shakiness during surgery. She later told me: "But I don't want to think of my baby's birth as traumatic." Our bodies and emotions tell one story, our lips tell another. Ultimately, as trauma expert Gabor Maté reminds us, trauma is in the eye of the beholder.

Aspects of my older daughter's birth felt traumatic—a nurse coercing me in a semi-drugged state to "just take the epidural" (contrary to my birth plan), having a second epidural administered when the first one failed, and then having an unplanned C-section after extensive labor. And yet it took me a year or two to actually associate that word "trauma" with my birth story. I remember feeling shocked to acknowledge that I'd experienced trauma during childbirth, and similarly disappointed by that label, as my client had been. But for me, it was important to acknowledge what had not felt right about the process. (Incidentally, I hear many stories of unwanted interventions and disappointing birth experiences, many examples of lower case and capital T traumas.)

Miscarriage, often a hushed and silent loss in our culture, occurs in roughly 20 percent of all pregnancies. Women suffer silently, often feeling alone and ashamed: *There's something wrong with me. What if I can't ever get pregnant? How come it came so easily for her?* If women could talk more openly about their own miscarriages and share their stories, it would help to normalize what is actually a common experience.

The false narrative that we tell ourselves is that if we don't talk about the hard stuff, it will make things easier. I'm here to tell you—both as a therapist and as someone who broke that habit long ago—it's total BS. We need to get real and talk about the hard stuff, too. That includes motherhood and fatherhood, parenting, the climate emergency, and all that we are facing today: the good, the bad, and the ugly.

We'll come back to trauma later, but for now let's remember that it doesn't dissipate like a puff of smoke once the problem is solved. Trauma lingers deep in the body's physiology, in the nervous system, in the neural circuitry, in the hormones, in the bloodstream, in the muscles and tissues, and in our cells. When we don't do the work of adequately healing our traumas, they come to haunt us in different ways—our physical or mental health, our relationships, our behaviors, and, yes, our parenting. Unresolved trauma can barricade us in so that we are unable to tend to others, whether it's children or the environmental crisis.

What's incredible is that we have all of the potential in the world to recover from our traumas if we are willing to face our demons, find healthy ways to release them, and be open-minded in the process. That might be through individual therapy, attending a support group, alternative medicine, creative expression, spirituality or religion, wilderness or ecotherapy, or just talking to the right person. Have you heard of post-traumatic growth? It's a real thing and a testament to our resilience, and it means that not only can we heal in the wake of trauma, but we can even grow from it as it stretches us in new ways. With time and adequate support, many folks find their way through pain and suffering, regain confidence, and experience resilience and greater self-awareness. Post-traumatic growth should give us hope in what's possible despite adversity. By integrating our pain into the narrative arc of our own lives, cultures, and ancestry—and witnessing pain as part of a collective—we can better heal ourselves for the work ahead. Moreover, as we shift toward understanding trauma as culturally and socially embedded—rather than as discrete individual experiences—we are more likely to enact a systemic makeover.

Straddling the Divide Between Vulnerability and Responsibility

But what about the sticky middle in between the highs and lows of parenthood (aka vulnerability)? Earlier in the chapter, I painted a picture of how tender the process of parenthood can be, stretching our mental health and physical bodies in new ways, and navigating so many unknowns. But once baby arrives, and we are suddenly—plop!—charged with the care of an innocent human life, fresh vulnerability surfaces. As the popular saying goes: "Making the decision to have a child is momentous. It is to decide forever to have your heart go walking around outside your body."

As I mentioned earlier, we can't underestimate parenthood as a story of grief and loss. While we gain the obvious—our beloved children, and heart-bursting love and devotion—we lose some of the security, confidence, and defense mechanisms we've built up along the way. But this molting is inherent in transformation, as we become like the Phoenix rising up from the ash.

By the time many of us become parents, to some degree we have self-organized and figured some shit out. For example, we have a sense of who we are, what we believe in, which relationships are important to us, our career, our lifestyle, hobbies, our preferred social media platforms, our favorite shows to stream, and so forth. When we become parents, all our life infrastructure—both internal and external—is suddenly challenged: our growing and evolving is happening alongside our own kids. And, yes, sometimes, like them, we regress, two steps forward, one step back.

Sticking with the lower case/upper case theme, some stories of parent *v*ulnerability include a dad feeling helpless and stressed after a layoff and prolonged unemployment; a mom smuggling in unapproved drugs from another country to help stimulate her breast milk production; or a child attacked by a dog with a minor injury. Some stories of parent *V*ulnerability include parents learning that their four-year-old child has a rare bone marrow disorder; a mom discovering that her ex was emotionally abusive toward her daughters; or a dad losing his wife to cancer while raising two young kids.

The inimitable Brene Brown has led a very important countercultural movement around shame and vulnerability. Through her research, she's determined that many of us avoid vulnerability at all costs because we associate it with heavy feelings like shame, sadness, and fear. On the flip side, she writes about its virtues: "Vulnerability is the birthplace of love, belonging, joy, courage, empathy, and creativity. . . . If we want greater clarity in our purpose or deeper or more meaningful spiritual lives, vulnerability is the path."[2] Whatever obstacles and pressures we may face as parents in today's world, if we can trust in our vulnerability as a source of strength and point of connection with others, then we'll gain an unexpected navigational tool.

Take a moment to find a comfortable position sitting, standing, or lying down. You might want to have a journal handy. As you consider the two reflection questions, pay attention to any sensations or movements in your body, posture, emotions, memories, and thoughts.

Reflection questions:
- What have been some of your big and small "V" vulnerable parent moments so far in this journey?
- What are some tender aspects of raising your kids in today's world or that you imagine will be tender in the future?

I work with a lot of teens and parents. Often the eldest child sparks emotional stirrings in a parent when they hit a particular milestone. When parents feel activated, stuck, rigid, or numb in their response to a child, I often ask them what that age was like for them. Not surprisingly, there are feelings and memories there, some of which they might be unconsciously projecting onto their kids. That's their cue to pause and do some of their own work so they can adjust to a more conscious parenting approach.

Parenthood

During my ten-year-old's life, various memories have flashed through my mind: slumber parties, crushes, school recess. Sometimes these are welcome, and sometimes they feel jarring. Memories will feel charged if there's an associated trauma. In an ideal world, parents are doing their own inner work as much as possible to help reduce their reactivity, defensiveness, or projections. Of course, we work with what we've got and where we're at, and come from a place of compassion (or "loving-kindness" in Buddhist-speak). The chances of getting it right every time? About zero percent.

In dominant American culture we've been conditioned to ignore our bodily sensations. But that doesn't mean we don't experience the magnetic pull of vulnerability in our heart space. In our quest to create a more just and honest world for our kids, it's time we reorient to vulnerability. Vulnerability is more than a personal North Star—it's a collective one for the Parent Club. Our pangs of vulnerability, of deep fragility, of intense fears and desires for our children keep whispering to many of us that we must do something, we must act, we must protect our kids, even if we don't fully grasp just how yet. Paying attention at this time is a critical first step.

And now for the second part of the equation. The one-two boxing jab that is the nature of parenting means that in addition to vulnerability you've also got a hot heaping of responsibility. Maybe you stop taking as many risks; you start working extra hours during the holiday season so you can save money for gifts; or your social calendar starts morphing into your kids' and your social groups start morphing into your kids' friends' parents.

Even as our feelings about our new responsibility may go haywire, we must rise to the occasion. We are no longer just living for ourselves; we have dependents, kin, and offspring to love, nurture, and raise to hopefully become content, well-adjusted, and self-actualized little beings. (No pressure!)

How do you reconcile these emotional dialectics of parenthood? Understanding the push-pull of vulnerability and responsibility is critical to being able to have both the strength and the emotional availability to discuss difficult global issues with your children.

The answer, I'm afraid, is the old Facebook favorite: "it's complicated." While this waxing and waning vulnerability may feel very uncomfortable at times, it's a critical part of being a parent—and, as we'll discuss later, it supercharges parents with fresh ambition and motivation to tackle everything from diaper changing to carbon drawdown.

I've heard many parents describe how surreal it feels leaving the hospital with their baby for the first time. There's a rhetorical question that pops up as a parent clicks their child into a car seat or stroller and prepares to leave a birth site: *Hold up! Am I really trusted to leave with this baby alone?!* More than one of my clients has done a maternity ward double take, looking back at their nurse or hospital room, baffled and confused that they are entrusted to look after their child without any help. Life is a gamble, but having kids increases the stakes, and we feel that deeply the moment our kids enter our lives.

One day, you are babysitting or playing with somebody else's kids (key words = somebody else's), and the next—you are the parent. This new reality can involve ongoing double takes, in which you look at your baby with a mix of shock, surprise, reassurance, joy, and apprehension. You might stare longingly at photographs from your freewheeling days of independence. Emotions will vary. Or you find that you're all absorbed and you don't look back until one day. . . . These double takes go on and on because your kids—surprise, surprise—keep a-growing! In fact, each of their developmental stages elicits a new emotional response from you (whether you know it or not).

To be fair, there are sweet moments when vulnerability and responsibility magically align, and it's like a huge sigh of relief. For example, when one of our kids is sick with a fever, my husband and I turn into the best versions of ourselves—doting, patient, and collaborative. It feels as though all the minutia is swept into tidy perspective. I know that I want to show up and be there for my child in that very minute, even if it means canceling plans or catching the bug myself. My love is amplified, and I find deep satisfaction in showing up as a responsible adult.

Parenthood

PARENT PAUSE

Shape-Shifting the Dialectic

- Take a moment to find a comfortable position either sitting or standing. You might want to have a journal handy. As you read the prompt that follows, pay attention to any sensations in your body, posture, emotions, memories, and thoughts.
- Think about a time in your life when you felt caught off guard, hurt, exposed, lost, confused, embarrassed, or ashamed. What did you instinctively long to do? Hide? Grow bigger? Hug someone? Withdraw like a turtle into its shell?
- Allow your body to move into a shape or posture to show how you felt in that moment. Feel into the shape and its sensations for as long as you'd like.
- Take a moment to reset afterward.
- Now think of a time when your child was threatened, hurt, bullied, injured, or ill, or you needed to advocate for them in some way. Try to think of a time you really showed up as a parent.
- Allow your body to move into a shape or posture to show how you felt in that moment.
- Feel into the shape and its sensations and hold it for as long as you'd like. Take a moment to reset afterward.
- If it feels okay to you, go ahead and bring your body back and forth, titrating between these two different shapes. How does it feel to move between them? Can you slow down the movement and pay attention to the in between?

- Try to find a shape that feels comfortable or sustainable between the two. Feel free to experiment. Breathe into this posture, a place where you can sense into both vulnerability and responsibility. Do any emotions, images, memories, or thoughts come up?

Note: If you feel triggered by this exercise in any way, please visit Appendix C: Tools for Managing Anxious or Traumatized Nervous Systems, for ways to help soothe and regulate yourself. Feel free to journal, make art, or share about your experience.

I want to acknowledge the work that you've done to this point! Dwelling on trauma, grief, vulnerability, and responsibility at the same time is certainly a tall order. Please take ample time for a break and reset before reading on.

Vulnerable Parents Raising Vulnerable Kids in a Vulnerable More-Than-Human World

It's a lot to hold all of this vulnerability in our hearts—kids and earthly life. You can keep practicing shape-shifting the dialectic titration (that's a mouthful!) as you read this section. Finding a sustainable posture that accesses your tenderness while supercharging your ambition is a reset to come back to again and again. The good news is that we can draw strength from the clear pathways and solutions mapped out by scientists, environmental authors, activists, policymakers, and others working at this intersection (more on this later).

As we flex our muscles of vulnerability and responsibility, we are preparing for the journey ahead. It will come as little surprise when I remind you that we're living through an era of postindustrial planetary collapse. Human-made outputs into the Earth's system—such as methane from factory farms, transportation pollutants, and oil spills—impacts the web of life in ways that have dire consequences for Gaia (Mother Earth). And Gaia, just to be clear, includes us humans. So, by extension, what is bad for the health of our planet is also bad for us.

Parenthood

Although Earth has self-regulating capacities, it also has its limits. There are sustainability tipping points. When we surpass these, we move into what is called "overshoot," a state that cannot be undone. Once polar ice caps have melted, we cannot "unmelt" them. There is no wand we can wave to repair the acidification of oceans or the destruction of coral reefs, and there's no easy fix for resurrecting razed rainforests. While there are certainly things we can do to heal, revitalize, and repair the Earth, planetary collapse is underway.

Scientists have adopted the term *anthropocene* (meaning "the age of human destruction") to describe our current reality in which human behaviors are a dominant face in shaping the Earth's systems. Like Mary Shelley's Dr. Frankenstein, we've created a monster that's gone rogue and is causing extreme harm and destruction. How can we possibly regain control and mitigate damage now that the human-caused monstrosity of climate change has begun to ravage the globe?

A major study published in *The Lancet* in 2021 interviewed ten thousand children and young people (ages 16–25) in ten countries (Australia, Brazil, Finland, France, India, Nigeria, Philippines, Portugal, the UK, and the US). Fifty-nine percent of respondents were "very or extremely worried" and 84 percent were at least "moderately worried" about climate change. Seventy-five percent thought "the future is frightening." Over 45 percent said their feelings about climate change negatively affected their daily life and functioning.[3] Closer to home, in 2024, *The Lancet* published a cross-sectional analysis of over 15,000 young Americans (ages 16–25) spanning all fifty states and Washington DC. Some of the results were strikingly similar: 85 percent endorsed feeling at least moderately worried about climate change impacts; 57.9 percent very or extremely worried; 42.8 percent indicated an impact of climate change on self-reported mental health; and 38.3 percent that their feelings about climate change negatively affect their daily life.[4] And in the adult world, according to Yale Program on Climate Change Communication's "Global Warming's Six Americas" study, over the past decade, the "alarmed" segment has grown more than any other audience (alarmed, concerned, cautious, disengaged, and/or dismissive), from 15 percent in 2013 to 28 percent in 2023.[5]

Children are particularly vulnerable to mental health issues related to the climate crisis. A Global Climate Risk Index by Germanwatch charts the extent to which specific countries have been affected by climate change (e.g., extreme weather events, air, water), showing that kids living in the Global South are the most negatively impacted.[6] Maternal exposure to pollution and wildfire smoke is associated with adverse birth outcomes, including pre-term birth and low birth weight. Data from Superstorm Sandy revealed that children in utero suffered from higher rates of anxiety than their nonexposed counterparts (53 percent compared to 22 percent).[7] A number of studies show children who survived the direct impact of hurricanes are at greater risk of developing sleep issues. When children experience environmental trauma or instability early in their lives, it can cause developmental trauma, PTSD, and other mental health issues.

Neurologically speaking, children's brains are major construction sites, developing in a bottom-up fashion. Higher cortical regions responsible for things like problem-solving, logic, and reasoning are last to develop. That means that seeing, hearing, or witnessing a disturbing event or news story will have a different effect on a child than an adult because they have not yet developed the neurological capacity for complex cognition, which supports emotional regulation.

In addition, kids haven't yet developed sophisticated defense mechanisms such as rationalization or compartmentalization to protect themselves from pain and suffering. While a kid's more open and less jaded attitude toward life is surely a strength, it also leaves them more vulnerable. Psychological distress often manifests somatically for kiddos, such as stomach aches, headaches, vomiting, insomnia, nightmares, lethargy, low appetite, or emotional eating.

Because kids watch and imitate other kids, one child's distress or concern can spread like wildfire to friends and peers. Suicide, self-harm, and violence can lead to copycat behaviors. With this in mind, you can imagine how dread can spread quickly throughout school communities, whether climate gloom or anxiety about school shootings. Raw emotions are amplified across group chats and social media platforms, increasing the risk of distress.

Author-educator Richard Louv warned us in his bestseller *The Last Child in the Woods: Saving Our Children from Nature-Deficit Disorder* that disconnection from nature harms children's mental health. He coined the term "nature-deficit disorder" to describe the human costs of alienation from nature: diminished use of the senses, attention difficulties, and higher rates of physical and emotional illnesses. He also pointed out the relationship between the absence or inaccessibility of city parks and higher rates of crime and depression. Anecdotally, many of us can recall experiencing a profound moment in the MHW—perhaps in a forest or observing an animal up close—that filled us with awe or thrilled or moved us.

Plenty of other cultures, particularly Indigenous cultures, live in right relation, or peaceful coexistence with the natural world. American naturalist E.O. Wilson coined the term "biophilia" in the 1980s to describe an innate affinity for nature and life shared by all of humanity. Ecofeminists have long drawn the connection between the oppression of women and subordination of nature, understanding how patriarchal structures exploit and undermine these two spheres through capitalism, colonialism, and extractive economies. For those cultures that have disconnected, it's catching up with us with a vengeance. As environmental destructions mount, rippling across our global communities, more of this disconnect is experienced, exposed, and forced upon vulnerable and Indigenous communities.

This human-environment disconnect has severe equity dimensions. Industrial pollution and waste facilities churn out unhealthy air, water, and soil in predominantly low-income and BIPOC communities that are written off as disposable, creating what are now called "sacrifice zones." Companies profit at the expense of human health and human lives. Environmental racism has serious cultural ramifications: huge segments of our population grow up in fear of the MHW, are more exposed to adverse health risks associated with environmental destruction (e.g., asthma, cancer), experience higher rates of depression that come from living in blighted landscapes with a lack of greenery, or are more likely to experience climate impacts more intensely. In America, stress compounds in structurally inequitable ways, making it very difficult to break out of the cycle. We must address these equity gaps in our policies, debates, and solutions.

According to a 2020 analysis by Conservation Science Partners, 76 percent of nonwhite, low-income individuals live in "nature-deprived" areas.[8] Green spaces—which include parks, hiking trails, community gardens, and cemeteries—are harder to come by in low-income areas. In addition to being beneficial for mental health, these natural areas act as cooling agents and sources of respite during heat waves. They can also mitigate short-lived climate pollutants. Children lose in these situations, particularly those in nonwhite, low-income families. And they are experiencing stark mental health consequences as a result.

In addition, large-scale climate catastrophes can lead to a cascade of negative effects in local communities such as increases in violence, abuse, substance abuse, suicidality, PTSD, and economic hardship. As climate impacts intensify, our connection to the natural world is threatened, human health worsens, and mental health problems multiply in a positive feedback loop of our own devising. According to a substantial body of research, hot temperatures negatively affect mental health, leading to increased aggression and irritability, deterioration in mood, and increased suicide rates. Expect to see more mental health fallout as a result of these temperature shifts.

All of this is a lot to worry about for the Parent Club. We must do better, and we can do better. Please take a moment for a break, some fresh air, and a reset.

The Parent Identity Spiral

Okay, reward time! When you're ready, look at the illustration that follows. It's an overview of the work that you'll be undertaking to shore up your identity, resilience, and strengths as an engaged parent doing their best to show up for your kids.

Parenthood

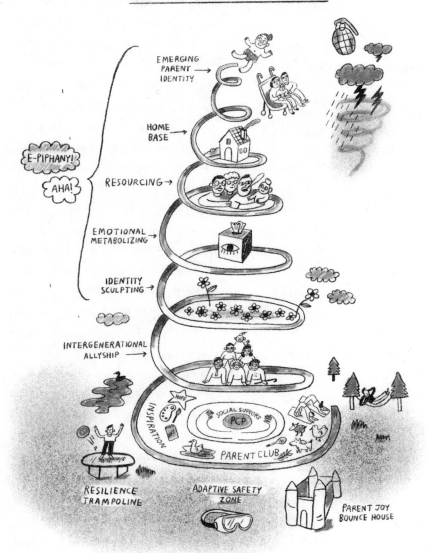

This illustration of the Parent Identity Spiral is, I admit, a lot to take in. But that's because it's a synthesis—and visual summary—of *all* of the inner and outer work that you're doing in this book. (Yes, this might very well be the page to dog-ear and return to if you feel lost at any point.) Note: A spiral was intentionally chosen since identity formation is neither linear nor one-size-fits-all.

- *Emerging Parent Identity* illustrates how parents are born—and start growing—alongside their kid's arrival. Raising kids can be viewed as a symbiotic process that encompasses new values, an evolving sense of identity, lifestyle changes, relationships changes, and work-life rebalancing, among other shifts.
- *Establishing a Secure Home Base* is the next phase of the spiral. In many cultures, parenthood involves nesting and creating a comfortable, safe home environment that corresponds with a vision or ideal of what a home should be. Of course, homes are embedded in what resources are available, your extent of privilege or marginalization, discrimination, or oppression. The notion of home is broader than just a place, and extends into emotional temperature, family dynamics, traditions, ability to adapt, and so on.
- *Resourcing* is your ongoing capacity to take care of yourself, lean on community supports, and ensuring that everyone in your family is resourced. It's a thread woven into the spiral, and reinforced by the trampoline below if all else fails. When you're feeling anchored and your cup is full, communicating with kids about tough real-world stuff will be easier.
- *Emotional Metabolizing* is your ongoing deeper reckoning with your emotions. By going deep into your emotions, it will be easier to hold space for your kids to process their emotions. It's a part of the spiral that you will inevitably return to again and again as emotions come and go.
- *Identity Sculpting* is the work that you and your family undergo as you seek a deeper integration between values, emotions, and actions, and start to imagine this in concrete ways.

- *E-piphany or Aha! Moment* is an awakening spark that may light up at any time along the parent journey and wake you up out of autopilot.
- *Intergenerational Allyship* is represented by a human pyramid in which different generations stand on each other's shoulders, holding each other up. The wisdom of older adults offers a strong foundation for parents (notably sandwiched) to work and support children from. Children have energy, freedom, fresh ideas, and rise up with promise. What a beautiful sight if we can get this right!
- *Social Support* forms a strong anchoring foundation, a place for you to access support, resource yourself, revitalize, and find joy and inspiration (i.e., the Parent Joy Bounce House, Resilience Trampoline, and Adaptive Safety Zone). I've provided several examples of what you might include here, but you'll develop your own personal concept of this throughout the book.

This Parent Identity Spiral is meant to offer you structure and guidance within a notoriously nebulous space. Structure, limits, and boundaries are all key aspects of parenting for kids and grown-ups alike that help to create a calming predictability, a sense of governance, if you will. Hopefully, this visual will be just that for you while also allowing plenty of room for personal interpretation.

CREATIVE EXERCISE

WHO AM I? PARENT COLLAGE

Materials:
- Paper, cardboard, or poster board of any size; you can also recycle by cutting up cereal boxes or boxes
- Magazines or pre-cut images, old calendars, photographs
- Scissors
- Glue stick, glue, Modge Podge, or tape
- Markers or pen

Setup:
- Divide the large paper (or cardboard or poster board) in two.
- Cut out images and glue them down when you are ready.

Prompt:
Arrange images on one side of the page that show what your life was like before kids. On the other side, arrange images that show what your life is like now as a parent.

Process:
As you create, notice without judgment, any feelings, thoughts, and sensations arising in your body.

Parenthood

Product:
When you're done, tape your work up on the wall and look it over with childlike curiosity, not judgment. Spend some time reflecting on themes, similarities, and differences, and consider a title for the work. Journal if you'd like to. Consider sharing your creative process and product with others.

Reflection Questions
- What was your life like before you became a parent?
- Who are you now as a parent?
- How have you changed or stayed the same in becoming a parent?
- What has this journey been like for you? Any ups, downs, or plateaus?
- What feelings have you experienced along the way? (e.g., Hopes? Fears? Sadness? Peacefulness? Doubts? Guilt? Excitement? Joy?)
- What or who are your sources of strength and support on this journey?

TWO

How Anxiety Operates in Families

"I wish my parents would stop micromanaging me so much."

–A teen client

As the first responders in our families, we are fated to bump up against external stressors, particularly those happening in our communities: hate crimes, violence, pollution, abuse, homelessness, and poverty, among others. I introduced the term polycrisis—the intersection of multiple overlapping crises that we face today—earlier, and I'll come back to this term. But first it's important to get a bird's-eye view of the stress that is bigger than just in our personal and family lives. I'll refer to the big world existential stressors as the "Macro." Such an intense backdrop while raising kids is notable, and can easily drive us into emotional overwhelm or avoidance.

On the other hand, there are the more personal day-to-day stressors that parents face in their homes and communities—financial stress, job loss, divorce, illness, or being evicted. I'll refer to this stress as the "Micro." By placing these two different sources of stress into two clear domains, it will help us to better grasp the nature of today's stress. Regardless of whether you're experiencing Micro- or Macro-related stress (or both!), how you act and respond hold major implications for everyone in the family.

In this chapter, we'll focus on how parent anxiety affects children. I've chosen to focus on anxiety rather than other forms of emotional distress because (1) anxiety in children is the most prevalent mental health issue, and (2) anxiety is a common and reasonable response in both young people and parents when facing both Macro and Micro stressors. According to a National Survey of Children's Health in 2023, over a five-year period the prevalence of diagnosed anxiety in children ages twelve to seventeen increased by 61 percent (from 10 percent to 16.1 percent).[1]

While I focus on climate anxiety in particular, the concepts in this chapter apply to *any* type of anxiety in a family system, whether it is Macro or Micro. Before we go any further, it's critical to acknowledge that climate emotions are a completely normal and rational response to human-caused environmental destruction. Those paying attention to the news, those directly impacted, those who feel intimately connected to the MHW, those studying or working at the environmental/climate intersection, those predisposed to worrying about the future, and those thinking about future generations are all going to feel some very big feels.

The last thing we need on top of our mounting existential distress is somebody telling us that there is something wrong about feeling this way—or something wrong with us! Climate-aware therapists commonly receive calls from potential clients because their therapists have disregarded or minimized their climate distress in some way, or told them that it's outside of their scope of practice. Luckily, these clients are able to recognize that something is off and will reach out for support, as plenty of therapists are still not yet thinking in climate-aware frames. As I will get to later in this book, climate emotions provide a rich opportunity to move out of paralysis into action.

Let's take a moment to distinguish more severe climate anxiety—one that impairs your daily functioning—from more benign worry. While some distress is incapacitating, other distress is what climate psychologists call "constructive" and motivates action. But whatever level of anxiety is relevant for you, this chapter will give you an idea about why too much of it, unfettered, can cause family dysfunction.

Before I continue, I'll clarify the difference between climate anxiety and eco-anxiety. Though frequently used interchangeably, they do have different connotations as well as some overlap. *Eco-anxiety,* short for ecological anxiety, refers to worry about our changing environment, including destruction of ecosystems, species decline, and shifting weather patterns related to rising global temperatures. *Climate anxiety* more specifically refers to anxiety related to climate change and the inherent chaos and breakdown that accompanies it, including intensifying climate catastrophes and mass climate migration.

Also notable is that I pluck anxiety from the larger soup of climate emotions. In chapter 5, I'll cover many of the other emotional variants that fall under the umbrella terms "eco-distress," "climate distress," "climate dread," and "climate emotions," so rest assured that there will be broader discussion outside of anxiety.

The last thing that we need in the midst of a complex crisis is confusing language. I get that. But keep in mind that climate psychology is an emergent field that is building the ship while inside of it. Language is shifting as the field grows.

The Trickle-Down Theory of Anxiety

As an intern trained in family systems and structural approaches, I learned early on that a child's distress is often the symptom of something much bigger happening in the family. You can't just assess a child, a dependent, in a vacuum. Often, if they are acting out, anxious, or depressed, it's because they are attempting to balance out an energy in the family system that they perceive as unhealthy or makes them uncomfortable, such as a parent who is drinking or yelling too much or invalidating their emotions.

A teenager at my clinic was referred for delinquent behavior. She was providing emotional support for her mother while her father was having an affair. She wasn't sleeping well because her mom crawled into her bed each night, crying, and seeking comfort. She felt worried about her mother, harbored anger toward her father, and missed having her own space to talk to friends and play music. Part of my work was resourcing her mother to find adequate emotional support so that her daughter wasn't shouldering as much of it.

In another case, I met with a senior in high school who carried the anxious aftereffects of her parents' messy separation. In a household where the adults took up a lot of emotional airtime, she learned to squash her own emotions, grin and bear it, and take on a mature adult role of peacemaking and caring for her sibling. In therapy, she learned to acknowledge her feelings, practiced healthy and assertive communication, and emotional disentanglement from her parent's ongoing tensions.

Since 2019, I've been wrestling around with climate questions from a family systems perspective. Some of the questions I mulled: Why were youth activists showing up in droves to advocate for climate action, while many grown-ups weren't? Was skipping school to attend climate marches really in a child's best interests? What did parents think about their child's level of involvement? Why was it so difficult for adults to safeguard precious environmental resources and ensure a healthy environment for future generations? As I wrote in a *Grist* article in 2021, "Instead of enjoying carefree childhood hobbies, they [young people] are protesting, litigating, organizing, and public speaking about the importance of policy-based climate solutions. Those efforts are courageous and inspirational, but teens' superhero-worthy feats do not absolve the rest of us."[2]

Where I've landed is that kids have adopted adult-like responsibilities in the climate movement to compensate for adult inaction and political gridlock. In therapy-speak, we say that kids who take on parent-level responsibilities that go beyond their stage of development are "parentified." Trying to save the planet from global environmental degradation qualifies as an adult responsibility, does it not? No matter what kids gain from activism, there's still a price they may unjustly pay for it, including missed school and academic lapses, unrealized hobbies, and missed social events. That's not fair to them, and it's a reminder that adults must do more.

As we can see by these examples, systems adapt under strain. In chapter 4, I'll get into some of the reasons why parents aren't talking about climate change or stepping up to the task at hand. But what is worth reiterating is that climate change is a systems-based problem that requires systems-level solutions. If we accept that climate change is a symptom of a dysfunctional global family system, then we must consider how powerful oil and gas executives and politicians who obstruct or delay climate change action—those in hierarchal positions of power—are a major source of the dysfunction. No amount of gaslighting should erase this basic truth.

The headline of my 2023 op-ed in the *San Francisco Chronicle* reads, "Call Government Climate Inaction What it Is: Child Abuse." By definition, child abuse and neglect is an act or a failure to act that places a child in harm's way. It can be argued that both child abuse and neglect are happening on the Macro level today. Failure to protect children from climate change has already led to significant loss of life, attachment ruptures, climate trauma, displacement and migration, and worsened health outcomes due to toxic exposures. Since there's no Child Protection Services on the global level, the best we can do is fire up our Parent Club (and advocacy organizations) to demand change.

Now, for a very important recipe that helps illustrate how climate anxiety can seep into consciousness and trickle down through your family. A friendly reminder: If you're anxious or worried about our climate and environmental crises, it is a perfectly normal, healthy response because it means that you are actually paying attention. Congratulations!

CLIMATE ANXIETY COCKTAIL

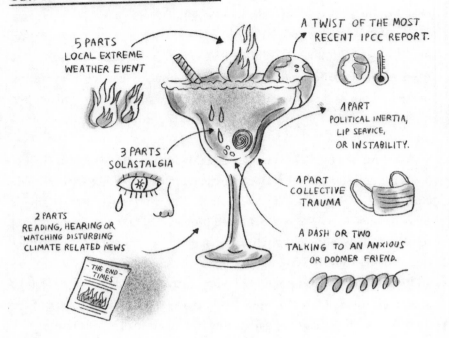

This recipe can be applied to any type of existential distress on the Macro level. It's also fair to use for Micro stressors as well. On the climate front, we are constantly bombarded by the news media with depressing statistics, disturbing images, or the latest death tally following a climate catastrophe. There's no doubt that absorbing all of this information in our minds and bodies is an assault on our nervous system and renders us feeling pretty helpless. The tension between all that needs to be done ASAP and all that is not being done by the powers that be is a perfect recipe for chronic anxiety.

Parenting can often feel like a ping-pong match. You do your best to try to juggle competing needs and demands to help keep everybody reasonably afloat. What can naturally occur as a result of this overstimulation is that you become out of sync with your inner landscape.

After a long day of work, chauffeuring kids to and from school and activities or scraping congealed pots and pans, parents deserve a break! Who could blame parents who might choose a hot shower or an episode of their fave binge show instead of opting in for the latest disturbing news?

We might work harder, avoid climate conversations that run too deep, immerse ourselves in hobbies or distractions, or squash the topic downward in the hopes it will magically disappear. But as psychologist Leslie Davenport told me in an interview for an article, those feelings can't stay down there forever: it's like trying to hold a ball underwater.

Let's not mince words either: we do not do our kids, future generations, or ourselves any favors by avoiding the climate emergency. Not only is living in a state of denial or disavowal problematic from an environmental standpoint, but it also creates unhealthy household dynamics. (More on that in chapter 4.) For now, let's consider trickle-down climate anxiety.

As parents, we might pat ourselves on the back for compartmentalizing our worries. But, in fact, opting not to talk about the climate crisis could very well send waves of fear rippling through ourselves and the family system. Kids pick up on our worry, then unconsciously or consciously think that it's off-limits to talk about climate, not wanting to poke the bear or feel that parents couldn't possibly understand their concerns since we never mention our own. Passivity, avoidance, and suppression—yes, these might be self-protective reflexes, but they can also be choices if we bring these patterns into consciousness.

A teenage client, for example, felt spooked when a science teacher told her class that in fifty years the area where they lived would be underwater. When my client approached her parents, they told her, "Oh, you shouldn't worry about that stuff at your age." Meanwhile, she imagines her family home and community drowning underwater, suffering in quiet distress. "I'm worried about what will happen to my family, so I try not to think about it," she told me. Like her, so many kids will learn to stuff it down, lacking outlets for these types of conversations, because it seems that few parents can bear to hear what kids are thinking or feeling about the climate emergency.

So how can we help to prepare younger generations rather than immobilize them?

A good rule of thumb is that unexpressed emotions will find a way to get expressed, but not always through words. Sometimes it may be through symptoms, behaviors, and negative thought patterns. I've seen anxiety in my clinical work manifest in so many ways: disordered eating, insomnia, substance abuse, self-harm, low self-esteem, controlling behaviors, acting out behaviors, interpersonal conflict . . . the list is endless.

Unconsciously or not, kids adopt their parents' emotional burdens in hopes of improving the situation. They might start mirroring some behaviors, acting out at school, or internalizing feelings to help balance the family system. Kids want to make the parents they love and care about feel better. But that's not their job—it's ours.

With climate anxiety and other emotions, your kids might notice before you do if you seem more anxious or depressed than usual. Perhaps you are spending more time alone, more time behind screens, yelling more often, or acting defensively. Some parents describe feeling numb, out-of-their-bodies, or have difficulty focusing on anything else.

A neighbor of mine with two young children told me that when wildfires are burning in the surrounding area, she checks air quality on an app obsessively, sniffs the air outside frequently, and analyzes the situation with her partner around the clock. She recognizes that her two young children, even if they don't fully understand the "why" of it, pick up on her anxiety.

I know a mother of three in Pennsylvania whose town was struck by unprecedented tornadoes, which sent her and her partner panicking, and plotting about how to create an emergency shelter below their house. You can bet their kids had an inkling that their parents were freaked out.

A colleague in Santa Rosa, California, a city that has endured multiple wildfires and evacuations, has noticed that dry, windy weather triggers both adults and children at outdoor gatherings, sometimes prompting them to move indoors because they are fearful after having lived through wildfires and associated trauma.

Kids not only detect changes in our moods (e.g., lethargic), in things we say (e.g., "the government is a mess"), or do (e.g., prepping for disaster) because they're hella smart and tuned in but because they feel the emotions circulating in the household deep within their bodies and souls. Chances are, you can bet that whatever the emotional tone of the household is regularly—worry, anger, sadness, love, peacefulness, hope—there is a slow drip down to your kids. At the end of the day, as much as possible, gatekeeping is part of parenting. Just as we protect an infant from a harsh environment, we must shield or filter out some of life's horrors from our kid's consciousness so as not to cause undue harm.

If you recognize yourself in the any of the patterns described above, there's hope—a lot of it. One of my favorite things about parenting is how forgiving it can be. We can shift course at any time, and kids are usually adaptive. Keep in mind that when we acknowledge our own uncomfortable potpourri of feelings, it increases our tolerance to hold space for our kids' feelings as well.

The parent relationship to climate anxiety also must include an on/off switch. Although we need to acknowledge the ginormous truth of the climate crisis and find real, active, and meaningful ways to engage with it, leaning into it at all times can be harmful and counterproductive. Our bodies, our minds, and our families require a state of rest (homeostasis) to support healthy functioning. Rest time supports immune function, gut health, sleep, and psychological resilience.

While the next chapter will delve into some actions that support life balance and resilience within families, we'll take a moment here to ground some of the anxious energy that might have just been stirred up.

Neuropsychiatrist Dr. Daniel Siegel and neuroscience leadership coach Dr. David Rock developed the concept of a "Healthy Mind Platter," a

The Healthy Mind Platter

The Healthy Mind Platter for Optimal Brain Matter

Copyright © 2011 David Rock and Daniel J. Siegel, M.D. All rights reserved.

wonderful visual that supports holistic well-being. As Siegel and Rock explain, "our mind . . . is in need of careful attention to establish and maintain mental health."[3]

As you can see, the Healthy Mind Platter has seven key components. While most of these categories are self-explanatory, it's worth some clarification: Time-In refers to quiet reflection, Focus Time refers to tasks and goals, and Downtime refers to relaxing (without an agenda). The Healthy Mind Platter is compared to a healthy nutritional diet, and its seven daily activities can be seen as mental nutrients that promote a healthy, balanced lifestyle. This visual can be hung on a fridge or bulletin board for easy reference. Whenever you or your child is feeling off, just visit this visual and reflect about what area needs attention.

What you do to repair, recharge, and care for yourself each day means everything. Parents are in the business of taking care of our kids for the long haul. Part of that is finding a way to sustainably engage in social movements. Two other foundational exercises worth attending to are (1) completing a self-care evaluation to get a handle on how you are doing, and (2) baking routine self-care activities into your weekly schedule.

PARENT PAUSE

Take a moment now to complete one, two, or all three of these exercises. Self-Care Evaluations can be found on my website or online, and the Weekly Self-Care Tracker can be found in Appendix B. For the Healthy Mind Platter, you can simply draw a circle on paper and divide it into seven parts that reflect how much time you typically spend doing each of the components.

Reflection questions:
- What are your strongest areas? Your weakest ones?
- Is there anything you'd like to do differently?
- Can you create a realistic plan for yourself?
- How can you hold yourself accountable?

It's all too easy to make excuses not to follow through. These tools will help with accountability. For everyone's sake, most of us need to turn off a worry switch well before bedtime each day. Some ways to create a transitional buffer before bedtime include setting a limit to your news intake (e.g., no tuning into the news after 8 p.m.), dimming the lights at a certain time, stretching your body before bed, and playing soft music.

I often suggest to clients to do a brain dump before going to bed, writing down whatever is on your mind without overthinking it. By doing so, you transfer some of your big emotions and nagging worries onto paper so that you don't have to take them to bed with you. You can (literally) rest assured that your thoughts won't be forgotten if you need or want to go back to them the next day. This ritual will help you bookend the day so that lingering thoughts are less likely to disturb or rouse you in the middle of the night. Try keeping a journal or notebook by your bedside for easy access. If you need to do this during the night to clear out your mind, you can do so with a booklight and not too much interruption.

Taking the time to rebalance ourselves each day is essential. And we should never feel guilty about that. In fact, modeling this for our children is wise.

A Word on Co-Regulation

As parents, we learn on the job about our own emotional reactivity. For someone like me, who once prided herself on not being a particularly moody individual, the emotional highs and lows I've experienced in motherhood are a tough pill to swallow. If we're not careful, a spilled glass of milk or a defiant remark can incense us with rage that doesn't match the transgression.

Observing when your nervous system is on high alert, or even better, catching the needle creep in that direction, is a crucial step in emotional regulation. Neuroscience explains how when we become dysregulated, our primitive reptilian brain structures take over. Suddenly, our cognitive functions are arrested. In fight, flight, or freeze, we can no longer think logically or act with reasonable consideration. And, of course, as we've now learned, dysregulation can be contagious: parent reactivity can ratchet up children's anxiety, which is the last thing we hope to do. (Don't forget, kids experience their own climate-related feelings to begin with.)

A tween client working on emotional regulation can teach us all a thing or two. They reflect on moments of their day when they felt upset or anxious, often drawing out what happened in pictures, telling stories about what happened, and how they felt inside. They describe a voice at an internal command center pushing a red alert button, an impulse to act, and how they are sometimes able to rescue themselves by taking deep breaths or ignoring a peer's provocations. This awareness is a life skill that is just as critical for parents as it is for kids.

The good news is if *we* can regulate our emotions, then our calming energy can also scale down our children's responses through a process called co-regulation. Therefore, soothing, or regulating, our own distressed nervous systems positively alters our children's physiology. Again, a good reminder that self-care practices are not selfish but are critical for parenting!

A General Theory of Love, by Thomas Lewis, Fari Amini, and Richard Lannon, introduces the concept of limbic resonance—that is, our nervous systems "speak" to each other because of interrelational permeability. Put plainly, limbic resonance means that humans are deeply affected by the brain chemistry and nervous systems of those around us. When we spend a lot of time or live in close proximity to someone, we attune to their biochemistry, and our limbic systems (the seat of our emotions) speak to one another. For example, if a parent's body is serving up a huge dose of dopamine (feel-good hormone) versus norepinephrine (stress hormone), it will shape a kid's emotional state, which over the long term can shape a kid's personality and even emotional health. If you can learn to regulate your emotions better, then you can hardwire that into your biochemistry and emanate less toxicity into the family-scape.

While studying with trauma expert Dr. Bruce Perry early in my career, I heard many disturbing case presentations about severely abused and traumatized children. Dr. Perry's calm voice inevitably came through the speakerphone as he would ask the presenter, "Did the child have access to a reliable, trusted caregiver or adult while this was going on?" If so, he told us, that would increase their chances of healthy development and recovery, because a secure attachment figure is a buffer against stress and trauma. Experiencing warmth, love, and calmness in a regulated adult figure firsthand not only supports co-regulation but it also is an opportunity for a healthy attachment blueprint (that will positively inform future relationships).

You might have noticed that when a child is upset, it can be helpful to lower your voice and speak softly, so that they almost have to lean in to hear. In this type of exchange, a child mirrors and in a sense borrows a parents' emotional capital, which helps restore their own. Co-regulation is a powerful tool and reminder that we can help soothe our children's distress. Oftentimes, a child's strong emotions (to put it blandly) are a test for parents. Kids are waiting to see how a parent will respond. *Will they tolerate their emotions, lose their shit and yell, or abandon them in some way?* Whatever happens, the main thing is to communicate to a child that you can handle the strong emotions that they are dishing out—and that they are valid (even if incredibly grueling).

If you do happen to lose your shit, never fear! You can always come back around and repair with an apology and explanation. Admitting to a wrong provides opportunity for forgiveness and fosters emotional connection. If you decide to give them space, you can let them know that you're available to talk once they've calmed down and are ready. Remember: Our attention is an expression of our love. That's why kids vie so hard for that reassurance each day.

A gift that we can receive, in turn, as parents is allowing kids to zap us back into the present moment, forcing us to toggle off. Kids want to be with us, play with us, laugh, make us laugh, push buttons, create, try new things, be outside, and run around naked (while shouting out inappropriate body part names). Guess what? We desperately need our own kids' wisdom here: There is a type of surrendering into kid mayhem that helps to reset us, a blessing in disguise. In the end, though, we sometimes have to humbly admit that our own kids did the co-regulating (or just helped to regulate *us*). We can aspire to do the same for them so that we don't inadvertently create an unhealthy dependency.

How to Manage Parent Worry Burdens

As a parent, there's a natural tendency to shield kids from the pain and hurt that we know exist in the world. Parents walk through each day holding the horrors of the world at bay so that we can offer our kids gifts of beauty, love, and hope. We aspire for a better world for our kids. This tension—and compartmentalization—can be a lot to carry.

From an evolutionary standpoint, our protective nature has served us well. But it doesn't always serve us well as individuals carrying these worry burdens around all of the time. There's a toll we pay when we don't manage our stress, and it will accumulate or persist. Studies indicate that some of the adverse health impacts of chronic stress include high blood pressure, diabetes, asthma, stomach issues, headaches, and skin conditions.

According to a 2021 NIH study, the US has one of the highest rates of parental burnout among the forty-two countries included in the analysis.[4] Since the COVID-19 pandemic, parental burnout is likely even higher. In an article in *The Atlantic,* physician Lucy McBride defined burnout as "the mental and physical fallout from accumulated stress in any sphere of life, whether that's work, parenting, caregiving, or managing chronic illness."[5] Of course, burnout can come from multiple directions at once.

As life rears its ugly challenges, and various events or news stories trigger our anxieties, we might try all manner of strategies to avoid suffering in an attempt to stay in the comfortable zone: using alcohol or drugs to unwind at the end of each day; overeating; working long hours to avoid going home; binge watching TV; engaging in hedonistic pursuits; and blaming others for our emotions or problems.

It's important to remember that people are doing their best to manage both the concrete and existential challenges that life poses. Societal issues, accompanied by a host of Micro stressors—such as birth or death, a separation or divorce, finances, illness, or an urgent community, school, or work issue—will not always come out on top for legitimate reasons. Any attempt to take care of oneself—perceived as healthy or not—is a form of coping. Of course, not all coping strategies are equal.

What coping strategies do you use to tame your emotions? How do you feel afterward, or the next day? Drinking alcohol or binge eating might provide an immediate sense of welcome relief, but the next day you might feel hungover, guilty, or depleted. With just a fifteen-minute walk, however, you can feel accomplished, reinvigorated, and less stressed. Your choices can make a difference.

Now, for a very important stress-relieving recipe.

Climate Anxiety Tonic

This tonic may come late in the chapter, but don't let that fool you: It's part of the resilience plan we'll be discussing more fully in the next chapter. The salted rim has an asterisk because it's the foil to the tonic that insists that we not forget the tough stuff. On the Jewish holiday of Passover, a day of reflective gratitude by way of symbolism, fresh parsley is dipped into saltwater. In Jewish tradition, the saltwater is a reminder of the tears shed by Israelite slaves fleeing Egypt long ago. Even as Passover is celebrated by kicking back and drinking some Manischewitz wine,

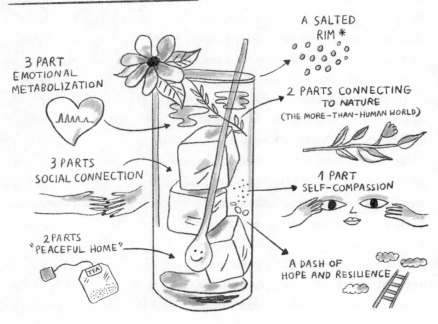

storytelling and symbolism remind Jews not to forget the bitterness of the past.

In the Climate Anxiety Tonic, the salted rim is a gentle reminder not to forget the bigger picture when we rest, pause, or retreat from climate engagement. Tasting the salt will help us to stay balanced and resolved, and not to slip into false optimism. Although climate anxiety isn't going away any time soon, there's at least a hangover remedy to help us out and buoy us when we are struggling.

A Word About Our American Family

While I'm not going to wade too far into political waters here, I think that it's worth applying some of these family system concepts on the societal level. US presidents are powerful leaders and, by extension, role models. We're living in unprecedented times as I write this in 2025. Donald Trump, unlike previous presidents, doesn't perform his duties with grace or humility or put the country first. Trump insults his opponents, hurls racist and misogynistic slurs, and pushes and breaks legal limits. If the head of our national household is unpredictable, narcissistic, lacks empathy, and puts himself above the law, then what message does that send our kids (trickling down, if you will)?

These messages reverberate within the walls of homes and communities. His administration has further blurred the line between objective reality and political interpretation by coining the idea of "alternative facts" to discredit and pave the way for propaganda. People have moved away from mutually trusted and established news sources toward ideological, radical news sources and social media, to the effect that Americans experience different interpretations of reality.

Trump's reelection in 2024 signaled just how lost and confused Americans are. As countless low-income and working-class Americans struggle with their daily lives, they pour false hope that an authoritarian figurehead will rescue them (i.e., rescue fantasy). And Trump inspires followers to scapegoat others as a punching bag for their anger (immigrants! Black and brown folks! Jews and Muslims! LGBTQ+!) as well as a return to traditional patriarchal (read: anti-feminist) values.

When politics get so bad that it costs people their own friends and family members, or it no longer feels safe to broach a heated subject at a dinner table, then you know a country is in deep doo-doo. That's America today. Politics has exacerbated class divides to such an extent that nobody feels safe. Gun safety measures—already weak and insufficient—are on the brink of further rollbacks. People are shooting other people who look suspicious, and police are grappling with unconscious—or even conscious—bias leading to untimely deaths for Black and other vulnerable Americans. Moreover, kids are shooting other kids and their teachers at their own schools. Shots are raining down every day while the rest of the world watches in horror.

By the end of 2023, the Gun Violence Archive reported 659 incidents of mass shootings in the US.[6] That figure doesn't even include single incidents of homicides, suicides, unintentional shootings, and police intervention and instances of brutality. The Gun Violence Archive defines a mass shooting as four or more people injured or killed in a single incident. That same year, there were 19,135 gun-related deaths (whether willful, malicious or accidental). Our country is an unofficial war zone.

When one such horror occurs, the nation grieves for a minute, and then one politician or another doubles down on blame-shifting to ensure that we can still buy guns. Second Amendment rights are prioritized over the safety of our own communities. As a sign that I saw outside someone's home states: "Protect our kids not our guns."

What are we to do in this culture but hug our kids tightly, perhaps sometimes overprotect? Even as we explore the anxiety within our families and homes, we must keep in mind the bigger picture. And learn from it. In the next chapter, you'll gain some strategies for managing a healthy home. But first, try this art exercise as a way to explore the flow of anxiety and the coping potential of tonic.

CREATIVE EXERCISE

CLIMATE RECIPES FOR RELIEF

Materials:
- 5" x 8" index cards (lined or blank), paper, cardboard or poster board cut to one size; you can also recycle by cutting up cereal boxes or boxes
- Markers, gel pens, pens

Setup:
Create two same-sized cards. If you do this exercise as a family, you can keep the card deck all together.

Prompt:
Consider working on this project as a family. Each of you will create a Climate Anxiety Cocktail recipe card that quantifies your triggers, as well as a Climate Anxiety Tonic recipe card that quantifies your self-care/coping strategies that you typically use in the face of stress (or can imagine using).

Process:
As you create, notice without judgment any feelings, thoughts, and sensations arising in your body.

Product:
When you're done, spend some time reflecting on your own recipes before coming together as a family, group, or with a partner to share what came up for you during this project.

Reflection Questions
- Which one of these two cards was the easier one to do? Why?
- Did anything surprise you or did you discover anything new about yourself?
- Do you feel that your tonic packs enough punch, or could you use some support in making it more potent? What might that look like? Just as with ordinary cooking, imagine how you might adjust or titrate these recipes to your taste and preferences.
- If you are comparing notes with others, are there similarities or differences among the different recipes? If you've done this as a family, consider any patterns in the family system.

THREE

Creating a Calm and Healthy Home

"The idealized version of 'home' means cuddling with my family on the couch, reading books together, feeling safe and cozy. . . . The reality of 'home' means listening to my daughter screaming all day, putting her in time-outs, and scolding my kids for fighting constantly."

—*Parent Club member*

For hundreds of years, humans have planned, created, destroyed, improvised, and reinvented the concept of homes within the context of our biosphere, interfacing in ways both sustainable and unsustainable. No matter what form it takes, home is at the very core of our existence. And yet we all know that our homes are more than just the structures we sleep in. The idea of "home" encompasses intangible feelings and visceral memories as well as sounds, voices, smells, and tastes.

Grappling with and determining what our homes mean to us, as parents, is fundamental—not only because we need to know what home is in order to provide one for our kids, but also because our homes are under both literal and metaphorical threat. Modern day threats come in many forms: from climate change impacts like floods, heat waves, and storms, to invisible threats like cyberbullying and catfishing, to discrimination and a pervasive sense of moral injury that others just do not care. This means that for those of us who've been so fortunate we cannot possibly take for granted the stability of our homes in the twenty-first century. Families will be challenged in new ways, as many parents will reckon with domestic uncertainty, learning to prepare for, create, recreate, and adapt healthy homes as circumstances in our communities change.

For many of us, home is rooted in the idea of a homeland, and our historical relationship to that land. We may experience a sense of safety through familiar habitats and ecosystems, like rivers, forests, trails, or beaches. If you probe deep, you might feel this, whether sitting along the banks of a nearby creek or climbing an old tree in a city park. Many people are fortunate to access positive memories by connecting to the world around them in a primal way, and these memories can offer light when we need it most. We'll explore our own ancestral relationships to the MHW through an art exercise later in the book.

A favorite art directive of mine is asking clients to draw or paint their "calm place." Whether a picture of a bedroom, a favorite vacation spot, or a particular landscape, I've witnessed clients smile, sigh, laugh, and stare at their artwork longingly. The calm place is a representation of home.

For those who've experienced forced migration, war, genocide, appropriation, or destruction of homelands firsthand—or who have been handed any of these as a legacy—home can conjure rage, injustice, grief, loss, and shattered stability. Home can become synonymous with trauma—a profoundly disturbing paradox.

Harm to landscapes, such as ecocide, polluted waterways, droughts, or floods, can trigger a cascade of consequences that impact homes for long stretches of time. For example, the degradation of a homeland, a farm, or a home can trigger food and financial insecurity, which can lead to health issues, malnutrition, financial strain, and stress.

Adapting our homes as external circumstances change is something that many of us recently lived through and experienced. During the COVID-19 pandemic, quarantines and stay-at-home orders redefined our lifestyles for a period of time. While incredibly difficult, with some families more hard-hit and vulnerable than others, our society has slowly recovered and continues to adapt as the months and years go on.

This chapter will look at seven different approaches to creating an adaptive home despite our unique family contexts, and very much changing, circumstances. But first, let's take a quick look at those Macro forces that come knocking on our doors, even looming inside our walls.

Meet Poly and Meta: Invisible Infiltrators

Before we set foot into solutions, I want to briefly address the unique modern-day threats that challenge today's concept of home: what are called the polycrisis (Poly) and the metacrisis (Meta). The polycrisis, as you may remember, is the intersection and overlapping of multiple crises at once. Poly says things like: "Environmental issues are human rights issues," "The hunger crisis is worsening due to climate change," or "War is ecocide." Poly generates a tremendous amount of existential angst that haunts parents' and kids' minds at night. We've learned to live under a pall of awareness of other's suffering, even as we're immersed in our own day-to-day lives.

Many families are directly mired in the polycrisis—particularly in the Global South, war-torn areas, and Indigenous communities. As economic independence, food insecurity, environmental destruction, and forced migration shape family life in a major way, homes are under direct threat. It's critical to reflect how some of us, as Parent Club members, are implicated in—or possibly buffered from—the polycrisis. When we are removed from its direct effects, it can feel like watching a movie unfold from a comfortable seat. But the suffering is real, and we must keep pinching ourselves to recognize the pain that is happening in many places at once.

The metacrisis is an internal crisis of despair and delusion arising from what Scottish philosopher Jonathan Rowson calls "the spiritual and material exhaustion of modernity."[1] Meta says things like: "I feel alone and hopeless" and "What's the point of doing anything?" The metacrisis is the psychological response that stems from the felt sense of disconnect in response to the larger systemic dissonance and violence playing out globally.

Poly and Meta are unwelcome lurkers in our homes. They can zap our energy, our hope, and our joy if we're not careful. But they also help us to understand the totality and connection between our inner and outer experiences. They are like twins germinating from the same place, reminders that we're living through a crisis of disconnect as parents raising kids in the world today—a disconnect from our bodies, from the MHW, and from other humans.

In her groundbreaking book *The Psychological Roots of the Climate Crisis*, British psychoanalyst Sally Weintrobe shares why it is that we have arrived at this point. She makes the case that 1980s neoliberal exceptionalism created a culture of "uncare"—a culture of hyperindividualism, consumerism, and inequality—that now pervades the world.[2]

A me-first culture embedded in relentlessly buying cheap stuff has deeply fragmented (and distracted) us at a time when we're desperate to come together, pool our strengths, and do something. Even as we tune into Netflix, online shop, post selfies on social media, and text about superficialities, our world is on fire (!), our sea levels are rising (!), our air is choked with carbon dioxide (!), war is killing innocent civilians (!), gun violence rages (!), hate crimes cause hurt (!), and racist tropes continue to fester overtly and covertly. Our distraction is no accident. When we're distracted, those in power stay in power—or may easily ascend into power—and people relinquish the power of democracy. Sound familiar?

By starting to care more and pay greater attention, we—as individuals, families, and communities—will help to grease the wheels of what

Buddhist eco-scholar Joanna Macy calls "The Great Turning" toward a life-sustaining society and the transition to a culture of care. Our youngest generations are already doing this. If we can revive a felt sense of connection, of human heart and care, then we'll find a way to rebalance. In the meantime, this chapter will equip you with ways to keep Poly and Meta at bay in your homes.

Seven Strategies for Building a Crisis-Resilient Home

Adapting to changing conditions has been an inherent part of a parent's duty for centuries—from wartime to post-wartime, immigration to forced migration, tourism to loss of habitat or lands, colonialization to segregation, "tough on crime" policing to gentrification and evictions. The list goes on, and will do so into the foreseeable because to live as a human is to adapt to whatever people, policies, politics, and environmental conditions come your way.

Wealth accumulation, financial security, and technology have created a privileged bubble granting a false sense of security and stability in life. As Poly heats up, and makes life less predictable for more and more people, these bubbles will pop. We need some guidelines for building adaptive, accessible forms of resilience. Lucky for you, that's what the next section is all about!

This chapter is devoted to examining your Climate Anxiety Tonic. Keep in mind that this is a tonic to buffer against any type of emotional distress in the home—Micro or Macro. There are seven strategies toward crafting this tonic, which I'll be covering in detail here: (1) rest, (2) parent centering practices (PCPs), (3) "peace of the home," (4) nature, (5) adaptive parenting, (6) deflating perfectionism, and (7) creativity. Throughout the book, you'll build on and replenish this tonic in new ways. This is only the first sip.

1. Getting Enough Rest

Rest may seem basic, but it's foundational. German psychoanalyst Erich Fromm once wrote: "The paradoxical situation with a vast number of people today is that they are half asleep when awake, and half-awake when asleep, or when they want to sleep."[3] I think about these wise words often. We naturally struggle against ourselves, and create our own suffering, yet his words suggest that if we can find a way to wholly commit to either enterprise—sleep or wakefulness—we can be more efficient, relaxed, and reclaim our time! (Loud cheer from parents everywhere.)

Physician Saundra Dalton-Smith offers a great TED talk about the seven types of rest that every person needs.[4] She parses out physical rest, mental rest, sensory rest, creative rest, emotional rest, social rest, and spiritual rest. It's important that we pause and take stock that *each* of these buckets is critical for our well-being and deserves consideration.

In mainstream America, most folks are certainly not getting enough sleep (more than a third of adults get less than seven hours of sleep each night according to a CDC 2014 study).[5] Lack of sleep comes with tough consequences: depressed mood, poor judgment, difficulty learning or retaining information, a compromised immune system, and long-term health consequences.

Our screen-based culture can interfere with sleep for adults and young people. Teens have difficulty separating from their phones at night, and younger kids turn stormy when parents attempt to separate them from videos or their tablets. I see tired faces and yawns in my office every day. A 2018 CDC study found that up to 73 percent of high school students are getting too little sleep.[6]

But before we even get started with our kids, we must first help ourselves. For every parent I see in my office complaining about kids addicted to screens, I hear a child accusing their parents of the same vice. Screens are highly addictive, releasing dopamine in the brain, and negatively affecting impulse control. In moderation, some screen use is fine, but too much is harmful for healthy development. Parents need to reclaim awareness of these tipping points and create rules and routines around them (e.g., one hour of screens on a school night).

For parents, there's a tendency to keep on going like the Energizer Bunny even outside of work hours, or when you say that you're going to take a rest or nap (*Let me just check one more thing on my phone*). I can attest to wanting to fill in every minute to stay on top of laundry, groceries, and bills. And with the 24/7 nature of social media and group chats, there's a constant pressure to keep up, keep checking, and don't miss out! Humans, wired to connect for their very survival, find it difficult to tune out of such pressures. It turns out that the very notion of rest these days is countercultural, privileged, and heavily judged—when, in reality, it ought to be a cultural prescription to benefit all of us equitably!

Dalton-Smith's other areas of rest are also critically important: sensory (being in the present moment attuned to your senses), creative (setting aside nonperformative downtime), and spiritual (feeling connected to the world in a way that moves or inspires you beyond your ego). I'll cover these in more detail in chapter 7.

PARENT PAUSE

Take a moment to consider *your* relationship with rest, and which of these seven areas might be ripe for growth.

2. The Parent Centering Practice (PCP): A Self-Care Building Block

For many parents, the day starts out with a bang, whether it be literally by a banging door, or by crying, an early wake-up, cobbling together a quick breakfast, or rushing everyone out the door. Another mama once confided, smiling, "My first cup of coffee in the morning is the best part of my day. After that, it's all downhill."

By the end of the day, things have only intensified: one hand is steaming broccoli, while the other is scrolling text messages. Pets and kids are clamoring, hungry, and tired. When the kids finally go to bed, there is a certain "blob-u-lar feeling"—a mixture of exhaustion, relief, eagerness, and confusion. Many parents are desperate to carpe diem because *shock* this is our only time to ourselves. This precious hour or two—or even three, if we're lucky—can feel like your only time to connect with a partner or yourself during the week.

And yet—what if you carved out a few minutes each day to check in with yourself?

As a term, "self-care" is so overused these days that it deserves a fork and some eye rolls. Nonetheless, the concept is still very important, and it is part of the framework for our resilience. A Parent Centering Practice (PCP) is a brief form of self-care that is specific to parents.

A PCP is an intentional act to calm or recalibrate yourself by connecting to the present moment. There are many ways to approach this and there is no one-size-fits-all. Parents desperately need to get a solid grasp on this to face up to our real-world challenges and traumas that pepper our lives today.

I'm definitely not the only parent at my school who squeezes their PCP exercise into morning drop-off as part of the weekly routine. Ideally, I work into these five to ten extra minutes time to meditate and set an intention for the day. Some parents build exercise into their commute by biking or walking to work. Not everyone can make morning the time for a thirty-minute-plus PCP, but how about a short meditation, stretching, or walking the dog? Or perhaps during a lunch break? After your kid's bedtime?

My daughter's piano teacher encourages students to practice their instruments even for just a few minutes of time—what he calls "exercise snacks." For example, he will practice chords on his guitar while waiting for the teakettle to boil. He says that this helps build up the muscle memory and creates good habits, and his own kids follow his lead. This is a useful approach for PCPs because it's a reminder that they can be brief and spontaneous, yet still effective, rippling out to the family.

A large portion of my work with clients is helping them to devise and carry through self-care strategies. At a basic level, this might be taking a few deep breaths every time they enter the car or observing rather than skipping a lunch break. A colleague of mine recalibrates between clients by mindfully washing her hands, a symbolic, tactile act that gives her a boost and reset. An art directive that I use frequently is writing or gluing an inspiring word/phrase/tea bag message onto a river rock. Some of my clients keep these stones inside of their desks as an anchor and reminder to look at during their workday.

My father-in-law, a clarinet player, has pointed out that what makes a piece of music resonant are the pauses, or the space between the notes. In other words, the rests between the notes are integral to the music itself. Similarly, the power of the pause is a gift to the score of parent life. I'll share some ideas for PCPs in a moment, but first let's talk about a simple, yet impactful one: deep belly breathing.

As numerous studies show, altering your breath quickly resets you by promoting relaxation and alertness, and reducing symptoms of depression and anxiety. Check out some of these techniques.

Kid-Friendly Exercises

- **Five-finger breathing.** Place both hands in front of you. Take one hand and start tracing the outside edge of your fingers starting on the pinky side of one hand, and tracing up one side and down the other side of each finger. Try to match this with your inhales and exhales so that one side of the finger is the inhale, the other the exhale (i.e., a full breath per finger). When you finish one hand, switch over to the other. Shake out your hands and wrists after and notice the aftereffects. You can do this two to three times.
- **Hot tea or hot cocoa breathing.** You can actually make some hot tea or cocoa or just use your imagination. Holding the mug, practice breathing in the smell and warmth and steam of the beverage, and

breathing out by blowing across the top of it to cool it. Try to use your senses to really feel into this. Repeat five to ten times.
- **Sillies shake out.** Yes, feel free to play the Raffi song on this theme, or your preferred song. But the main point is to shake, stomp, jump, shriek, laugh, skip, hop, dance, or what have you as you let your body take over and feel into that (be sure to quiet that inner critic).

Slightly More Adult Exercises

- **Simple orienting.** You can sit or stand for this. Look around and find something pleasant to look at while focusing on breathing in and out. Notice the details of what you are looking at and feel free to think or say them aloud. Keep tuning back into your breath and then back out to observation.
- **Alternate nostril breathing (a yogic breath practice called "Nadi Shodhana").** Tuck the second and third fingers of your right hand so that you have access to your thumb and ring finger. Inhale slowly and bring the thumb of your right hand to your right nostril and gently close it while exhaling slowly out of the left nostril. Then inhale slowly in the left nostril and at the top of the inhale place your ring finger on your left nostril to close it, exhaling through your right. Inhale through the right and cover up that nostril to exhale out of the left. Repeat three times on each side.
- **Stretch and flow breathing.** Follow your body's intuition to combine stretching and breathing, paying attention to what parts of your body want attention and movement. Hold and breathe into stretches to allow the release of tension. This can be sitting or standing or a combination—you can add music, a yoga mat, or do this outside. Feel free to follow a yoga practice or find a video online if that's helpful. Make it your own!

These breathing exercises are just a sampling to get you practicing breath connection more regularly. Meditation apps such as Insight Timer, Calm, or Headspace offer guided meditations of varying lengths and can be a great place to start because they provide structure and direction. The same would apply to joining a weekly meditation group. Keep in mind that deep, conscious breathing is a practice that takes time to get the hang of. Practicing these exercises regularly helps to restore parasympathetic function to your nervous system. Yes, your mind will vie for your attention, and in many cases win. The trick is noticing when you devolve into what is called "monkey mind" and just refocus with gentle compassion, much like you would guide a puppy away from gnawing on your shoes.

Trial and error will teach you what works best for you, and how much time is necessary. At minimum, five to ten minutes for this once or twice a day, or if possible, a good solid twenty- to sixty-minute chunk of time. By reconnecting with body and breath, parents move out of autopilot toward a healthy reset. At the same time, kids can practice ways to deescalate anxiety before it blooms into panic. The rhythm of breath, and its depth and flow, is foundational for a healthy home. (See "PCPs to Help Reset You During the Day" for additional ideas.)

PARENT CENTERING PRACTICES

PCPS TO HELP RESET YOU DURING THE DAY

Listening to or playing music
Warm shower or bath
Walking
Dancing
Singing
Your preferred form of exercise
Hugging someone (or even yourself)
Talking to a friend
Crafting
Sunbathing
Spending time in MHW
Relaxing in a hammock
Gardening
Cooking/baking

If you haven't already, be sure to fill out the Self-Care Weekly Tracker in Appendix B. Embedding PCPs into your weekly routine will help you in the long run.

Creating a sense of safety, a place for you to land first and foremost, will allow you to feel more spacious, and better meet your children's needs. When we go through difficult times, our feelings can unintentionally transfer to our children; there is only so much containment that we can realistically do. With great intention then (and many colorful sticky notes), we have to figure out how to keep filling our cups, feeding our emotional piggy banks, and reaching for our oxygen masks. Even when we feel stressed to the max there is often some wiggle room somewhere to do *something*; there is choice.

3. "Peace of the Home"

When I spoke with pediatrician Dr. Kenneth Ginsburg about ways that parents can manage existential stress, he articulated a concept that I'm not only familiar with but also hold dear as a parent: maintaining a sense of harmony at home regardless of what's happening in your internal and external life. In Hebrew, the phrase *shalom bayit* means "peace of the home," and describes the Jewish value of domestic harmony, originally associated with marital relations. The idea is to hold the value of harmony high as best you can at all times. This concept is a good reminder for families to maintain a loving home with care, compassion, and respect.

Living with so much uncertainty is tough. With all of the very real threats on the Macro level waxing and waning as part of the new normal, it can be difficult to unwind at the end of each day, even when we are safe. While hypervigilance might help us to survive and prepare for an uncertain world, it's not healthy on a cellular level, contributing to chronic stress and health conditions such as chronic lung, heart, autoimmune, and gastrointestinal diseases.

I think of a former client—he wasn't more than seventeen years old—who was trying to leave his neighborhood gang. He and his girlfriend had recently had a baby. As a community mental health worker, I met with him at his family's house in a quiet residential neighborhood. During our sessions, whenever a car drove by, his back tensed, his ears perked, and he'd abruptly cut off conversation. Sometimes, he'd stand up and peek through the side of the curtains to make sure he was safe, and that a gang member wasn't looking for him.

When you live in a community or home plagued by violence, you can't take your safety for granted. Your system adopts hypervigilance as a default mode, continually scanning the environment for any signs of threat (i.e., what is called "neuroception"). This constant hypervigilance takes a toll on your physical and mental health—and on your family members. Sometimes, my client's baby cried from the next room. He'd walk over to her with a blank face, and peer at her, emotionally disconnected. How can we possibly meet our child's needs when our own systems are so dysregulated? The answer is we can't, or we will miss some things, try as we might.

While the concept of peace of the home provides a helpful template for most families, it will be much more difficult to institute in some circumstances. If you're currently living in a home where you feel reasonably safe and secure, then peace of the home can be of great value, helping your family to feel anchored despite the uncertainty outside of our doors. When so much is unpredictable in the world, intentionally creating safe havens in our homes, as Dr. Ginsburg reminded me, supports us in feeling calmer and gentler toward others. There are many ways to do this.

In my own family, we follow what we call "Slow Sundays." These are a weekly reminder to preserve the flow of spontaneous, unstructured life: to spend time in our garden and with one another, to kiss expectations and efficiency goodbye, and to see what emerges. I've come to observe my partner smiling more than usual, my girls giggling while playing school in our garden, while I take up residency in our hammock as some of the most pleasurable moments of family life.

I will admit that in the frenzy of living in an urban environment, the advent of the smart phone, and the business of my kids' social and activity calendars, this relaxed atmosphere has been harder to cultivate. However, it always remains a moving target that I try to re-create in my own home despite changing conditions.

Creating a Calm and Healthy Home

PARENT PAUSE

Reflection questions:
- In what ways is your home peaceful?
- What are some of the obstacles to making your home peaceful?
- Now consider your family of origin and/or pre-kid life. Are there any traditions or rituals that you would like to integrate into your current family life?
- How might you create a peaceful home during uncertain times?

To help you on this quest, check out this list:

Peace of the Home Brainstorm

- Power down devices at certain times of day (place out of sight and on silent mode)
- Regular family meals
- Family bonding activities (e.g., games, sports, crafts, cooking, yard work, cleaning, trips)
- Religious or spiritual traditions (e.g., shabbat, solstice, holidays, prayer, singing, meditation)
- Daily quiet/down time
- Sensory modulation: dim lights, quiet time, candles, smells, tastes
- Keep the peace in front of the kids, argue and work out relational kinks on your own time
- Date nights
- Spending time with relatives and extended family
- Decluttering

- Simplifying routines and expectations
- An abundance rather than a scarcity mindset (i.e., we have enough, gratitude)
- Repeating household mantra or sayings (e.g., when I was growing up, returning from a trip, my dad would shout out as we entered the house: "Travel east, travel west, in the end home is best")

I'm going to focus now on families in transition. Wherever you are in the world, under many circumstances, it's possible to conceptualize and execute a home base appropriate to your context—even if circumstances are temporary and far from ideal. There are so many unhoused families in this country, living transiently. People in families with children had the largest single year increase in homelessness between 2023 and 2024, with 39 percent more experiencing homelessness.[7]

Over the years, I've worked in various transitional living homes: family shelters, domestic violence shelters, foster homes, transitional residences for young adults, and juvenile hall. Many folks max out their time and have to move on to new placements, continually uprooting their family and belongings. It's incredibly difficult, and painful—and particularly hard on kids. The suggestions that follow are intended to support immigrants, refugees, climate refugees, families struggling in the wake of a traumatic event, families who've lost homes, transient or unhoused families, anyone moving or relocating with frequency, or someone in similar circumstances.

How to Work Toward Peace of the Home Amid Familial Transition or Displacement

- Keep safety as the number one priority
- Look for trusted adults, allies, or "helpers" (in *Daniel Tiger's Neighborhood* terms)
- Ask for help, and teach your kids to do so as well

- Create or join communities that feel inclusive and supportive
- Practice daily or weekly rituals as much as possible: morning kiss, note in lunch box, bedtime story, gratitude prayers at meals, reflections at end of day
- Keepsake box of memories (e.g., photographs on a table or the fridge, memorabilia/objects, or a shoebox of letters)
- Reset safety and evacuation plans as needed so kids know the new program
- Talk to kids in appropriate language that will help calm them during stressful circumstances: "We are going on an unexpected adventure, and it might feel hard at times, and not as comfortable as before, but we will be together and others will help us."
- Prioritize rest when possible
- Lots of hugging, affection, and extra soothing
- Singing familiar songs, humming, listening to familiar music
- Eating comfort foods

4. A Nurture Nature Family Philosophy

As many of us intuitively know, and as research now corroborates, spending time in the MHW improves our mental, physical, and spiritual well-being. In Japan, *shinrin-yoku*, or forest bathing, is a mainstream medical intervention prescribed to reduce depression and anxiety, lower blood pressure, and boost immune function. Studies indicate that gardening, or horticultural therapy, is also therapeutic. Soils contain bacteria that can trigger the release of serotonin, functioning similarly to an antidepressant. Interacting with animals has also been shown to decrease cortisol and reduce blood pressure. Many of us have experienced these benefits in our lives firsthand.

Why do we not do this more often when we know how crucial it is for our well-being? The answer is the standard one: We get time stressed. We feel like we just can't fit it in. (And, of course, there are equity dimensions too.)

How do you appreciate the natural beauty around you? Does your family swim in lakes, rivers, or oceans in the summer, play in the snow or go sledding in the winter, go on hikes when the leaves change in the fall, or welcome the first signs of spring when flower buds emerge, and mammals mate and give birth and tend to their young? Is the climate mild all year long, as it is for me in California, and seasonal changes are less obvious? If you are living in an urban environment, what are some ways you connect with nature? Perhaps you own a pet, hang an indoor plant, plant veggies on your rooftop, have a favorite local park you spend time in, or support a community-supported agriculture project (CSA) or natural foods co-op. How does your family relate to the MHW? Is there anything you'd like to add or change? How did your family of origin relate to the MHW when you were growing up? How might you apply knowledge from that experience?

As a bonus, spending time in nature with kids can be like seeing through new eyes. Kids are naturally attuned and enlivened by their five senses. They notice things grown-ups don't. My youngest daughter frequently reminds us to "look at the beautiful sunset!" when we are driving home at the end of the day. This way of being can help tamp down the frothy babble in our own heads, reminding us to do the same.

The chances are that if you are reading this book you have some interest or curiosity in the MHW already. And, as happens with parent passions and interests, these trickle down to your family. (Of course, they can also reverse flow like the Nile and trickle upward from kids to adults.) In fact, studies indicate a correlation between time spent in nature and pro-environmental behaviors. There are ways in which many of us are already doing this, and ways in which we can do more, appreciate more, and reflect.

To get you started on this endeavor, here is a checklist to help you think about this more in depth. Following the list is a Parent Pause that can also be for the whole family.

Creating a Calm and Healthy Home

Family Eco-Values Checklist

- ☐ Taking care of animals and/or the environment
- ☐ Eating healthily and/or sustainably
- ☐ Living close to nature
- ☐ Spending time in nature (e.g., hiking, camping, picnicking)
- ☐ Gardening/planting/weeding
- ☐ Volunteering
- ☐ Helping others develop a relationship to nature
- ☐ Attending climate groups/meetings/events
- ☐ Vegan diet
- ☐ Vegetarian diet
- ☐ Pescatarian diet
- ☐ Reduced meat/free-range diet
- ☐ Mindful usage of electricity and water
- ☐ Transitioning gas appliances to electric
- ☐ Solar panel installation
- ☐ Walking, biking, skating, or taking public transit instead of driving
- ☐ Buying used clothing
- ☐ Limiting plane travel
- ☐ Recycling
- ☐ Composting
- ☐ Gardening
- ☐ Supporting a CSA
- ☐ Using a reusable coffee mug
- ☐ Bringing grocery bags
- ☐ Refilling at bulk stores
- ☐ Buying refill bottles
- ☐ Not using straws or using reusable straws
- ☐ Boycotting palm oil unless sustainable
- ☐ Buying recycled products
- ☐ Buying cruelty-free products

- ☐ Teaching others about caring for the environment
- ☐ Sharing flowers or goods growing in your garden with others
- ☐ Taking care of animals or pets
- ☐ Weeding invasive plants
- ☐ Planting trees
- ☐ Combining errands to reduce emissions
- ☐ Buying sustainable fish or free-range meat
- ☐ Buying local, humane certified, or free-range eggs
- ☐ Practicing "Leave No Trace"
- ☐ Boycotting oil-divested companies
- ☐ Buying local and organic
- ☐ T-shirt or bumper sticker activism
- ☐ Other: _____
- ☐ Other: _____
- ☐ Other: _____

PARENT PAUSE

Take some time to complete the Family Eco-Values Checklist on your own or together with your family. There's room to add your own ideas and practices to this list, too. If you are doing it on your own, consider checking in with an accountability partner. If you are doing it as a family, you can complete it together. This can be posted on a bulletin board or fridge.

Reflection questions:
- What values are your family currently practicing?
- What values would you like your family to start practicing?

- What values and priorities are most important to each of you?
- Are there any sticking points that are difficult to negotiate? How might you devise a shared sense of ecological values?

Creating an eco-values list can help to ground you when you are feeling anxious about *anything*. It can be a more general values list, certainly. Living your values—truly embodying them rather than operating via lip service—brings a sense of purpose and relief that counters anxiety, overwhelm, numbing, and despair.

One of my favorite expressions from graduate school is "work smarter, not harder." Although not a multitasker by nature, I do think it's possible to achieve multiple goals with some forethought. As a parent concerned about the climate future, there's a way to prioritize the MHW, the climate crisis, family, and your own self-care, all at the same time. Approaching climate engagement as a family bonding enterprise and embedding it into your core values, rather than keeping it a solo mission, connects the dots between mental health, climate engagement, and parenting.

5. Adaptive Parenting

While lying alone on a beach in Deer Isle, Maine, during a summer trip with my family, I stared in a semi-zoned-out way at the offshore white buoys that bobbed around Penobscot Bay. Boats entered the harbor to dock, others set out for sea, winds changed course, and the tide ebbed, but through it all those buoys remained, winking at me in the late afternoon sun, unfazed.

If only I could be like that buoy, I thought. *Whatever is thrown at me, I'm okay, I can come back to surface, and bob through choppy waters.* Buoys are man-made devices, little aquatic helpers designed to guide ships, boats, and people to safe harbor. But more than that—they adapt to changeable conditions. If a buoy sinks underneath a boat, it will quickly pop up again. Nothing can bring it down. Buoyancy—resilience—is in its design.

The question is: How can parents buoy their families through unchartered waters, even as they are knocked about, battered, or pulled under?

While it ain't easy-peasy, it's certainly possible. We can guide our kids to safe passage, build stronger communities, surf emotional waves big and small, and act as a beacon of hope, like a lighthouse blinking through fog. (For that matter, there's even a parenting style called "Lighthouse Parenting" that you can check out.) Our ability as humans to adapt to changing conditions—to buoy up—is one of our biggest strengths. (Nautical metaphors work well here and water happens to be the most adaptive of all elements, so go figure.)

Buddhist scholar Pema Chödrön writes: "To be fully alive, fully human, and completely awake is to be continually thrown out of the nest. To live fully is to be always in no-man's-land, to experience each moment as completely new and fresh. To live is to be willing to die over and over again."[8] Adaptive parenting is a response to Chödrön's insight: work *with* rather than *against* present conditions. To show up, to feel, to acknowledge, and to respond. Not to squash, ignore, minimize, belittle, control, or cling. And yet, it's freakin' hard.

Whatever is happening to us or around us, we have a choice in how we respond. Do we yell at our child when we are at the end of our rope, or take a loooooong breath in and out (a PCP!) and place a quiet arm around their heaving shoulders? Do we scroll the newsfeed under the counter while our child is telling us about their day, or put away our device to listen better?

In Dialectical Behavioral Therapy (DBT), a common evidence-based mental health treatment, we have three choices to any situation: we can change it, accept it, or suffer. Usually when I tell clients this, I get a smile. It really boils down your discomfort to a three-pronged decision. And who wants to choose suffering?

In a way, the term "adaptive parenting" is redundant. Isn't all parenting adaptive? Kids develop, life changes and evolves, we age, get sick, get better, and so on. Those with less privilege are forced to adapt more than those with more privilege. Money can insulate us against change: anti-aging regimens, relocating or traveling, or buying gadgets to feel safer or more comfortable. Those with fewer means are inevitably more vulnerable, with fewer opportunities. For example, housing projects are often in urban areas with industry, pollution, and lacking green spaces; low-cost housing might be in coastal, low-lying, or high-risk wildfire areas and more prone to climate catastrophes. Vulnerability is a function of privilege that we cannot deny.

In the wake of the 2020 global pandemic, we've learned a good deal about adaptive parenting—and even developed some coping strategies as life as we knew it ceased to exist. Think about what a radical transformation transpired in just a couple of years! The world looked and felt different because business as usual came to a squeaky halt. Psychologist Shelley E. Taylor described humans' innate tendency to come together to socially engage amid stressful events as a "tend and befriend" instinct.[9] In other words, under threat or stress, we nurture our young, and seek out social support and resources. This is critical for our very survival.

PARENT PAUSE

Recall those early days during the COVID-19 pandemic in 2020.
- How did you initially feel about the pandemic? How did you react?
- What are some of the ways in which your family adapted?
- Was there anything that you resisted?
- Can you identify a positive or negative moment from that time? How did you carry on?

Note: If you feel triggered by this exercise, please visit Appendix C.

This exercise offers you an opportunity to reflect on how you or your family has adapted to changing life circumstances. A client of mine, a mom, once shared with me some advice she received from her midwife as her labor started to progress: "Honey, the contractions are gonna happen whether you breathe through them or whether you pound the floor." The point is that you're not in control of what happens to you but you do have control in how you respond. Bracing or steeling yourself can sometimes make things worse. You don't want to lock down into tense rigidity; you want to stay lithe, flexible, and as calm as possible.

6. Deflating Perfectionism with a Pop!

As I write this section about adaptive ninja-like parenting, a caveat: Beware of perfectionist tendencies! Not only is perfectionism an unrealistic ideal, but it will also do harm to you and your family. In my therapy office, as of late, I've been hearing teenage girls lament how they are "not good at school" or feel like "a failure" because they don't have straight A's. They feel a tremendous amount of pressure—from their parents, their teachers, their peers, and social media—to excel at just about *everything*. This is absolute crazy-making, and not a healthy setup for our kids, to say the least!

In the therapy room, I help both adults and children examine harsh beliefs about self and others while helping them challenge and reframe these in compassionate ways. Instead of "I'm a failure because I suck at math," for example, a reframe might be, "I'm having a hard time in math right now, but I'm feeling good about my work in history." I call these compassionate reframes, and they are helpful in reinventing our toxic mental scripts.

I could write a whole section about perfectionism and motherhood dedicated to how moms strive toward perfectionist ideals and feel ashamed when they cannot meet them. Social media assaults moms with curated photos or videos, which are eaten up like Lay's potato chips with little questioning. *Look at that mom's hot postpartum body! Look at how happy she is as a mom! Look at how happy her baby or family is! She's such a strong mama because she had a "natural" birth. She's already back to work?* And so on. I will say that I much prefer a candid shot of a mom friend biking with her family, ruddy-faced, sans makeup and all smiles, than the posed seasonal shots of another mom friend all made-up in formal attire and photoshopped to meet some perfect family ideal.

I also want to make a plug for healthy modeling here. If I work with a youth client and they start doing the work of dismantling their own negative self-talk and replacing it with healthy, strength-based self-talk, then kudos to them! But what happens when they return home and hear from their parents: "You should do this or that . . . are you even trying?" or "You're fine . . . don't cry . . . you're going to be fine." While parents often have good intentions, sometimes messaging goes awry, or they will jump to problem-solving rather than connecting or just listening.

A parent dismissing or ignoring a child's feelings is significant. Children absolutely notice, and will feel sad, ashamed, anxious, humiliated, and guilty (fill in the appropriate Mad Libs® adjective). Like most of us, children cling fast to criticism, or fill in the blanks with negative interpretations. Let's not forget that we all have a negativity bias—a tendency to allow negative comments or life events to affect our well-being more than positive ones. "I'm not good enough," is one of the most common negative cognitions, and it really tanks self-esteem. At the end of the day, compassion and understanding go further than harsh decrees, strong-arming, and unsolicited problem-solving.

When parenting in supercharged times, self-compassion is absolutely key, as you are going to try out new things and stumble at times, sometimes with your kids watching. The last thing anyone needs—you or your family—is your own condemnation when you are already trying hard enough. Sometimes self-compassion needs a boost, like attention or intentional action. We can't just assume it will be there for us. Radical acts of self-compassion might be setting healthy boundaries or limits around your time, listening to a guided meditation, or just doing something kind for yourself that brings you simple joy.

7. Creativity as a Muse

Often, when I give a spiel about art therapy to a client, I'm met with one of these responses:

> "I've never been good at art."
> "I'm not artistic."
> "I'm not sure it's for me."
> "I had a teacher who told me ____."

Though I understand these sentiments, I also find them utterly sad because I feel that many folks have been robbed of their creativity early on in their lives. These voices echo a broader creative trauma that runs deep in American culture today. As a result, I spend a lot of time preparing clients in advance to keep an open mind. I remind them that this is not art class and you don't need to have any particular skills or knowledge to be creative. I remind them that art is just fodder for therapy to help process and understand deep feelings. As I overheard a colleague point out, many of us cook but aren't contestants on *Top Chef*. Many of us enjoy hiking, but don't worry about summiting the world's highest peaks. So why is art-making any different? Many of my clients can move past their initial defenses and give it a try, but not all of them. It's a shame that one of our innate gifts as humans—our creativity—is stifled early on by critical words. The only thing getting in the way of experiencing pleasure, rapture, and joy in the creative process is how judgmental we are of ourselves—and of each other.

Here's the thing: Creativity is (and I refuse to be modest here) a human superpower! It's one that can be harnessed by families facing modern-day existential stressors. Embracing the more intuitive, creative, and spontaneous side of human nature allows kids to feel deeply into their imaginal and unconscious realms, giving themselves permission to just be and explore. It's a true gift. We can support this indoors and outdoors, at home, on vacations, and anywhere we go, really. It's not limited only to visual arts but extends into building, inventing, exploring, writing, dreaming, cooking, singing, making music, acting, pretending, dancing, traveling, body movement, games, and so on.

Since creativity is under siege early on, it takes intention for parents to nurture it as a value. When my older daughter tells me she is bored, I feel slightly irritated. I think that what she is really voicing is that there's free time and she's not exactly sure what to do with it. I've often thought to myself *boredom is a failure of the imagination*. But perhaps a positive reframe of this sentiment is *boredom is a gateway to imagination*. In these moments, I gently help my daughter to brainstorm some options. She usually figures out her next move quickly.

An overreliance on screens as a solution to boredom—which I see all of the time—is a slippery slope. In some situations, it might be a quick fix that makes sense. But if parents automatically direct kids toward a screen every time they're bored, then they are doing them a developmental disservice. It's like Pavlovian conditioning, paring boredom with a screen. Sometimes I understand that it can't be helped, but other times there are different choices we can make.

When creativity bubbles up like a spring, it uplifts spirits in the household. Kids and adults feel that creative energy flow deep within; they connect with their own life force. Exercising creativity nurtures growth, collaboration, independence, perseverance, flexibility, resilience, and open-mindedness. As I see it, it's a fundamental aspect of a healthy home.

In the face of climate change and other real-world challenges, creativity and adaptability go hand in hand. As we face new challenges and are pushed out of our comfort zones, we can start to imagine, innovate, and spark new solutions. This can affirm hope and optimism, an anti-doomer mindset. When we view life rigidly and as unchangeable, it's easy to fall prey to a hopeless, doomsday mindset.

Creativity offers us a possibility to see things from a fresh perspective, to find meaning in hardship. "When life hands you lemons, make lemonade" is an example of asserting our creativity and literally transforming something bitter into something sweet rather than succumbing to despair.

Children are curious agents in the world—less jaded than many adults—and are positively brimming with fresh ways of seeing things, new inventions, and questions of all kinds. One of the worst things we can do as parents is snuff out this innate curiosity and extinguish their creative spark, which is a gift to this world.

As this chapter draws to a close, you'll do an art activity to help you synthesize some of these seven aspects of a healthy home and develop your own personal definition of what that means to you. Not all of this will resonate with you, and that's okay. You're invited to create your own understanding here of home.

CREATIVE EXERCISE

HOME SWEET HOME

Materials:
- **Scrap wood**
- **Shoe box, tissue box, or other recycled box**

- Any found objects from the natural world or from around the house (e.g., fabric scraps, cotton balls, tissue paper, stones, branches, flowers)
- Clay, modeling clay, popsicle sticks, toothpicks, bottle caps
- Hot glue or tacky glue, collage images, paint supplies
- Sensory items (e.g., sachets, herbs, sprays, edible, tactile, bells, photos)

Setup:
Gather materials and set up your studio. Dive in!

Prompt:
Create a 3D representation of a home that is meaningful to you. Consider both the interior and exterior of your box. This representation can be abstract, conceptual, literal, representational, imaginary, sensory, populated, and so on. (If you liked the boat and buoy metaphor, then consider creating a boat, and even an anchor.)

Process:
As you create, notice without judgment any feelings, thoughts, and sensations arising in your body. Reminder: this is a curious exploration, not an art class!

Reflection Questions
- What do you notice about the exterior of your home? The interior? Are there any themes?
- What mood does your home elicit in you?
- What threats exist outside?
- Would you like to give your home a title?
- Are there any questions that arise for you as you sit with your home?

FOUR

Why Parents Bury Their Heads in the Sand—and How We Can Look Up

"I try to pretend like wildfires aren't happening."
—Parent Club member

One of the advantages to writing this book is that I'm an intended audience member as well as a clinician-writer called in to creatively problem-solve. In other words, I've got some serious personal intel on this situation. So when I tell you that I understand why it's difficult to pay attention and sustain attention to all the depressing stuff going on in the world today, I really do get it from the inside out. Just to further prove my parent cred, check out the abridged list in "20 More Enticing Things to Do Than Face the Climate Emergency as a Parent."

20 MORE ENTICING THINGS TO DO THAN FACE THE CLIMATE EMERGENCY AS A PARENT

1. Build a Little Free Library.
2. Vacuum the stale debris in and around the car seat.
3. Draw up or revise your will.
4. Color-code your kids' books.
5. Replace an air filter.
6. Hose out your compost and recycling bins.
7. Sort the Tupperware drawer.
8. Take your kid anywhere.
9. Use a Q-tip. Thrice. On each ear.
10. Ask this question of ChatGPT: Is ChatGPT worth it for humans?
11. Take a deep dive into your astrological moons and houses.
12. Find somebody else as excited about enneagrams to talk "wings."
13. Start working on a Shutterfly photo project.
14. Update the privacy settings on all of your social media platforms.
15. Take a CBD gummy and see what happens.
16. Help your kids' set up a lemonade stand in the neighborhood. Clean up.
17. Start a PhD program.
18. Teach your child a lanyard stitch.
19. Clean out that junk drawer.
20. Respond to neighborhood watch threads on Nextdoor.

Why Parents Bury Their Heads in the Sand—and How We Can Look Up

Jill Kubit, cofounder of Our Kids' Climate and DearTomorrow, told me that parents are "uniquely motivated" to do something about our climate emergency. After all, parents want nothing more than to support their kids the best way they can and yearn for them to succeed and find happiness.

At the end of the day, it's our kids who are going to inherit the polycrisis. Our role is to prepare and empower them. When we pay attention or take action, we not only embed these values into the fabric of our family culture and communities, and into the hearts of our children, but we send our kids a clear message: *I got you.* One day they'll do the same for their kids.

This chapter will explore some of the psychological barriers to Parent Club climate engagement and provide you with some concrete tools so that you can take small steps forward on what I will unveil as the Yellow Brick Road of Climate Behavioral Change. While this chapter is climate-focused, the ideas can apply to most stressors, addictions, or behavioral issues.

Parents Are Overloaded Yet "Uniquely Motivated" to Care About Climate

Parents of every generation have set out to do what they are biologically programmed to do: raise offspring against the odds, often in hostile environments. Whether the threats coming at them are wildebeests, tribal warfare, religious persecution, natural disasters, famines, wars, slavery, or climate chaos, parents throughout the millennia have contended with adversity while managing to survive and raise healthy children. Through the twentieth and twenty-first centuries, each generation has faced a set of defining societal challenges: polio outbreak, women's suffrage, World War II, Vietnam, Civil Rights Movement, *Roe v. Wade*, LGBTQ+ rights, Black Lives Matter, #MeToo, conflict in the Middle East.

What's on the docket for today? Yes, you know it; we've talked about it: a "code red" climate emergency, with a side of political obstruction. The challenge today is fighting powerful entities hell-bent on preserving the fossil fuel/money pipeline. We are up against Hobbesian forces: money, power, violence, and greed. This "shadow side" of humanity, as Swiss psychiatrist Carl Jung refers to it, has been stoked by unbridled capitalism and the backlash of its inherent inequities.

The climate crisis has existed for so long in a nebulous space, looming in and out of our consciousness depending on what scientists are saying, the media is covering, or the politics du jour. In fact, the climate crisis has inhabited this grey space for so damn long that we've grown a little too comfortable letting it remain there. The social movement that we all need to join right now is accepting human complicity—and therefore responsibility—in the climate chaos engulfing us so that we resolve to face it squarely, collectively, and effectively.

Intersecting with climate change are other Macro-level concerns lurking, including war, poverty, human rights, racism. The existential laundry list can be quite daunting.

Not surprisingly, studies indicate that anxiety is soaring among expecting, new, and seasoned parents, and parental burnout rates are higher than ever. Many parents—disproportionately women—carry both a high mental load (read: running a household, work stress, caregiving) and a high emotional labor load (read: absorbing and managing emotional dynamics in the family and community, sandwich generation). These heavy loads are compounded further by structural failures, such as a lack of federally protected paid family leave, lack of affordable, high-quality childcare, and lack of affordable, high-quality healthcare. Our European counterparts know what's up and have long established policies that uplift parents, children, and families. According to a 2019 Organization for Economic Cooperation and Development study, the US is the only country out of forty-one that lacks paid parental leave.[1]

Notably, since the 1990s, there's also been a cultural shift toward hyperintensive, down 'n' dirty parenting styles in the US, skewing toward time-consuming, stressful, and competitive. You've likely heard some of the labels for today's parents: "helicopter," "snowplow," and "tiger." The names alone are quite telling, but for a rough definition: helicopter = hovering, snowplow = micromanaging, and tiger = achievement pushing. No rest for the weary parents.

In light of the sheer business and stressfulness of modern parenting, it seems quite adaptive to avoid thinking about existential threats that *may* harm your family at some point in the future. But, as we know, it's also deeply problematic because we can't rely on the powers that be to do the right thing.

At the end of the day, it's part of our collective responsibility as parents to ensure our kids have a livable planet, to be stewards, caretakers, and good ancestors. Cultural anthropologists have described ancestors working together on child-rearing tasks as "alloparents." Some examples of alloparent culture exist today in Indigenous communities in the Republic of the Congo and in the Philippines. Considering the degree of mind-boggling systemic failing we are up against—governments, institutions, businesses, and compromised Conference of the Parties—we can't afford to stay stuck in our own emotional muck. Our kids' futures, the futures of our grandchildren, and the health of our living Earth are all at risk.

Hello, Defense Mechanisms, and Other Barriers!

Climate change has long been tough to grasp and hold onto, even when we are paying attention. Philosopher Timothy Morton first introduced the concept of "hyperobject" in 2010, suggesting that climate change, which spans both time and geography, is too big and abstract to be grasped in its entirety; therefore, it's easily ignored. Not only has this played out over decades through skepticism of climate science, politicizing of climate science education in school, and difficulty altering our own carbon footprints, but it has done so as well despite reports by the Intergovernmental Panel on Climate Change (IPCC), increasing media coverage, and a strengthened global climate movement. The hyperobject looms over us. Indeed, many live by the credo of Adam McKay's 2021 film: *Don't Look Up.*

It's also important to point out the how fossil fuel companies intentionally manipulated consumers for decades to protect their own selfish interests. British Petroleum (BP) hired a PR firm in the 1980s to promote the idea that climate change is the fault of individuals rather than the fossil fuel industry. It was BP that popularized the phrase "carbon footprint" and created a "carbon footprint calculator," and environmentalists and others ate all of this guilt up with a spoon. (Not that they are at all to blame for what is deceitful, harmful behavior on BP's part.) The book *Merchants of Doubt: How a Handful of Scientists Obscured the Truth on Issues from Tobacco Smoke to Global Warming* (2010) chronicles how politically conservative scientists joined forces with think tanks and corporations to deliberately sow doubt and confusion about climate change in an effort to obstruct progress. Indeed, there's plenty in the climate landscape that has fueled and continues to fuel avoidance, even before we dive into the depths of the human unconscious mind.

Let's take a look now at some of the psychological defense mechanisms at play. Some of these might ring a bell, others you might relate to or see in other people.

Evolutionary Psychology

"Run!"

"Don't move a muscle!"

"Quick, hide!"

Reactivity is the name of the game here. For the sake of our very own survival, humans have evolved to prioritize immediate threats first—like the saber-toothed tiger chasing you, the rising floodwaters outside your door, or gunshots ringing in your neighborhood. Our brains are *not* wired to respond in the same fashion to a gradually warming climate, or the effects of El Niño impacting a faraway part of the world. Existential threats like climate change or structural racism, while they are toxic crises that require our immediate attention, are abstract until they act or register on a visceral level—involving your community or life directly. Over time, we just get used to them being around, and we move into passive acceptance.

So what do we do about it? It means that we must find ways to attach a sticky note to our frontal lobes with a list of what we conceive of as truly urgent threats. It means paying attention even when it's a beautiful day out, you're on your summer vacation, or its offseason for hurricane or wildfire threats, yadda yadda. In short, it means climate gets triaged regularly.

Stress Response System
"OMG!"

When you hear "climate chaos," "breakdown," "collapse," "emergency," or watch disturbing footage, your heart beats quickly, you might feel panicky or floaty, and disconnected from your body. It just feels like too much, so you try to avoid thinking about it. Only it's getting harder and harder to avoid because climate change is all over the news now. Our sympathetic nervous system's fight/flight/freeze/submit/fawn stress response is extremely uncomfortable and debilitating to stay in for too long. But that's not the only thing. If we get stuck in a cycle of chronic stress, it harms our physical and mental health long term and it will hinder us from taking meaningful action. There are ways to regulate your emotions and find a calmer window so that you can tolerate and weather the extremes. We'll dig deeper on that in chapter 6.

Denial

"I try not to think about it."
"If I don't think about it, then I won't worry so much."
"Scientists will figure it out."

While writing this chapter, I spent a week with my seventy-seven-year-old dad, who had been under the weather for over two weeks with a low-grade fever, aches, and lethargy, but kept dismissing the need to check in with his doctor or set up a phone or medical appointment. "I'm fine, I'm fine, I'm feeling better now," he kept saying whenever I prodded. "Did you take any pain medication?" I learned to ask. "Well, yeah, I did about an hour or two ago," he'd respond in his sheepish manner. "Well then," I said, "you're not better yet."

The point of the story here is that denial comes in many forms, including minimization and rationalization. Here are some common ways denial shows up in climate conversations:

> *"Well, that stuff is happening far away—it's safe here."*
> *"What can I possibly do about it?"*
> *"What's the point? We're all gonna die soon."*

Denial, as most of us will admit, will only get you so far. You unconsciously or semiconsciously repress difficult truths and emotions. Somehow, it's never a good time to unpack your distress, or allow feelings about heavy topics to surface. Like a dam building up pressure over time, negative emotions build within and can lead to serious health issues like hypertension, gastrointestinal diseases, depression, panic attacks, or suicidal thoughts. It takes conscious energy to suppress unwelcome emotions. That energy is wasted and unnecessarily exerted! Instead, just imagine if that same energy was directed toward productive, fruitful endeavors such as climate conversations or political activism. Eventually you have to face the music.

The phrase "soft denial" is a good one to know, and you might even relate. *Soft denial* means that you acknowledge the existence of climate change, but this knowledge remains intellectualized and unexamined—at arm's length. In parenting terms, I think of it as how parents understand a lot of things in theory (e.g., the importance of healthy boundaries, not yelling at your kids), but often this doesn't translate into practice.

Disavowal

*"I know it's happening, but there are other things
I need to focus on right now."*

When I asked a mom if she felt anxious about her kids, she retorted: "Am I anxious? No, not at all. I don't do anxiety. I mean—what's the point?" I had to laugh. Is anxiety a choice? That's a good one to chew on for a while. To me, her words register as a type of disavowal (a close cousin of soft denial).

With disavowal, you understand on some level what is happening . . . but then you quickly distract or turn your back. In *The Psychological Roots of the Climate Crisis*, British psychoanalyst Sally Weintrobe describes it as a simultaneous knowing and not knowing of painful realities, in which the mind avoids the deeper emotional processing that comes by acknowledging the truth.[2] Disavowal can be quite reflexive, like a subconscious flicker. One moment, you're watching a news report of an intense flood and feeling intense concern . . . but then you pop over to your social media feed rather than go down an emotional rabbit hole or resolve to engage.

I've noticed many folks living in wildfire-prone regions tend to disavow in the offseason. After wildfires abate and air quality restores, there's a collective sigh of relief (at least, by those fortunate to escape direct impacts) . . . followed by a moving on, or tuning out, until conditions for wildfires intensify again. Disavowal is an evolutionary switch that turns on and off to protect us from prolonged states of arousal and chronic stress. While it's true that we need these buttons to protect our mental health, it's also critical that we do not dwell there permanently.

Psychic Numbing

"Say what?"

Psychic numbing can occur individually or collectively. A person who has endured a traumatic event might respond by detaching from others and showing little interest in activities, even with a restricted affect. Their immense pain guides them to self-protect. Renowned psychiatrist Robert J. Lifton has applied psychic numbing at scale to describe how an entire culture will withdraw from an issue that feels too painful to think about. Whether the bombing of Hiroshima or climate collapse, there's a human tendency to retreat or numb ourselves.

Trauma, USA

"I just can't go there right now."

Like many mental health interns acquiring client hours for licensure, I began my work in what are sometimes referred to in my field as "the trenches"—community battlefields, if you will. These are deeply traumatized, marginalized, and underserved communities struggling to make it under the mantel of structural racism, poverty, and other forms of oppression. I worked with survivors of severe childhood abuse and neglect, domestic violence, and sexual assault, with immigrant newcomers, with youth on probation or in the juvenile justice system . . . you get the idea.

Being a white therapist from a privileged background, it was profoundly humbling, eye-opening, and soul crushing to drop into what felt like an underground matrix of trauma, grief, abuse, and terror. But it really showed me the magnitude and frequency of suffering endured by so many. As I heard horror stories again and again, I learned quickly how trauma is baked into poverty and just how common a denominator it is in the human experience. Its prevalence, and its impact on so many innocent families trying to get by, made me livid. A voice within me screamed: look at how rampant the abuse in this country is! You cannot look away from this!

There are four main types of trauma: acute, chronic, complex/relational, and developmental. The groundbreaking Adverse Childhood Experiences (ACEs) study conducted by Kaiser Permanente between 1995–1997 surveyed over seventeen thousand adults regarding their childhood experiences and current health status and behaviors.[3] ACEs include physical and emotional abuse, neglect, domestic violence, and caregiver mental illness. The results of this study revealed a stark positive correlation: the more ACEs a person has, the more likely they will suffer from health outcomes like heart disease, diabetes, depression, substance abuse, poor academic achievement, and early death. Close to 64 percent of American adults report experiencing at least one ACE; 16 percent report experiencing four or more ACEs.[4]

This finding is important because it (1) illuminates how deeply traumatized the majority of Americans are, and (2) shows how early childhood trauma predicts negative health outcomes later in life. Experiences such as intimate partner violence, parental separation, and household substance abuse are traumatizing experiences for children and can shape their developing attitudes and worldviews, moods and behaviors, and health. Trauma symptoms often linger throughout their lives into adulthood.

Many adults who come to therapy are nursing childhood wounds and traumas. These can easily inform their parenting. Think about that for a moment: Many adults are yoked to childhood wounds and burdens that they've barely processed or broached. What might these unconscious wounds do? Well, they might lead a person to turn to alcohol or drugs, displace their pain onto someone through violence, blame, yelling, or criticism, or attempt suicide. This is the darker story of modern America.

In *The Myth of Normal*, psychiatrist Gabor Maté boldly details how trauma is baked into our culture and imposes a worldview tinged with pain, fear, and suspicion: a lens that both distorts and determines our view of how things are. This perspective is important to understand as trauma creates additional barriers to facing the kinds of big-picture challenges I discuss in this book. At a fundamental level, trauma disconnects us from ourselves, which makes it difficult to connect with others, or with the MHW—let alone be able to tolerate distress or regulate emotions so as to engage with world problems. Moreover, trauma robs us of our adaptability as we get locked into rigid and reflexive response patterns.

Personal Greenwashing (aka Rationalization)
"Look, I recycle... other people can worry about the big stuff."

I firmly believe that any type of contribution you make to protect the MHW or to combat climate change is a worthy contribution. To show up fully in your humanness matters. I've already said that individual actions count. So does self-compassion and acting in accordance with your values. However, to push forward the societal changes that we need to see on the big scale, and at the strong pace required, we must build a robust climate movement that includes everyone.

In my view, one of the engines could very well be a Parent Club full of people who care deeply about future generations, whose worries keep them up at night, and whose love could be fruitfully directed to make a difference. It takes hearts, minds, and bodies coming together to accomplish a herculean task such as this one. Child-rearing is the perfect prerequisite for this climactic moment (no pun intended); we have all of the experience that we need to act courageously.

As we move along, however, we must do our best to question the contradictions, defensiveness, and unhealthy tendencies we see in ourselves and resolve not to kick the can any further down the road. On an individual level, "Well, I'm doing X, so I can do Y" is not enough momentum. If someone trying to quit cigarettes resolves to visit an oxygen bar twice a week while continuing their habit, that's hardly productive.

Alas, I cannot leave the "Personal Greenwashing" section without acknowledging corporate greenwashing. Typically, this term refers to the corporate manipulation of consumers by exaggerated claims of environmentally aligned products, services, or policies.

But what about net zero initiatives? The pledges that some companies are making to strike a balance between greenhouse gas emissions and the emissions they remove from the atmosphere through carbon offsets or carbon removal.

While I'm not weighing in on this per se, it's worth thinking about how this paradox works against us: It's fine to pump more greenhouse gases into our rapidly warming atmosphere because at least we're cleaning up after ourselves! This hardly moves the dial toward a greener, cleaner, and more equitable economic system. It's a little like binge eating unhealthy food, but justifying it because, hey, you are exercising. These bargains with the devil, so to speak, are stalling out more aggressive and creative problem-solving along the green growth lines.

On an individual level, a harm reduction approach such as making eco-conscious consumer choices is worthwhile. But the systemic changes we need to see cannot be so incremental; they must boldly integrate economic, sustainability, and environmental justice initiatives into a concrete vision. In the haunting words of youth climate activist Greta Thunberg, "Our house is on fire!" A crisis of this level requires a time-sensitive crisis-level response, and the very best leadership we've got. President Franklin Delano Roosevelt's enactment of the New Deal is an example of bold and swift federal leadership that succeeded in lifting people out of the depths of the grim Great Depression. This is the type of societal transformation and trust in moral leadership that we are so desperate for now.

Cognitive Dissonance

"This doesn't feel right."

Cognitive dissonance is a fancy psych term for having conflicting values, beliefs, feelings, and behaviors. For example, when someone is pressured to harm another person—such as bullying someone at school or torturing a prisoner of war—these antisocial behaviors counter our prosocial programming. It's a feeling that something is wrong or "off," perhaps morally wrong or contradictory in a way that distresses you.

In a 2019 *Washington Post* article,[5] I shared a personal example of cognitive dissonance as a mom going through the usual motions of readying and dropping my kids off at school despite worrying about their safety due to wildfire smoke and nearby fires. I distinctly remember scrambling eggs on the range, and talking to my kids in as normal a voice as I could muster, while my stomach ached and I had a nagging sensation that maybe sending them to school was the wrong thing to do.

Following the 2021 Oxford High School shooting in Michigan, one parent on social media network asked, "Did anyone else hug their child tighter this morning?" Posted another: "I was crying at the bus stop when the bus pulled away this morning." How paradoxical it is that schools—educational institutions to support children's growth and development—are now community targets for random acts of violence?

In 2021, after the Uvalde and El Paso shootings had already rocked the state of Texas, leaving twenty-one and twenty-three people dead respectively, the Texas legislature passed a bill allowing Texans to openly carry guns without permits. This is a form of cognitive dissonance that is of political nature; an unsettling doubling down that uses a tragic event as the impetus for furthering a political agenda. Like rubbing salt into an open wound, moments like this one can intensely resonate because cold-blooded policymaking contradicts what's in our hearts. As parents, we feel this deeply. The Parent Pause that follows will help you to examine cognitive dissonance from the inside out.

PARENT PAUSE

Choose a topic that you have strong feelings about (e.g., LGBTQ+ rights, women's rights, the war in the Middle East). Grab a sheet of paper and a pen or pencil, title the page with your topic of choice, and draw four columns: Beliefs, Dissonance, Actions, and Obstacles. List your beliefs about the topic in the left column, any signs of cognitive dissonance in the next column, any actions that you've taken or that you would ideally like to take, and your barriers to taking action in the right column.

Reflection questions:
- Are you living your values and beliefs?
- How do you know if you are aligned—or not?
- Do you have a vision for aligning your emotions with action?
- Do any of the obstacles have potential workarounds?

Starting to examine our own defense mechanisms at work is one way to breed a sense of agency and to move through "stuckness" toward action.

Despair

On the subject of despair, I turn to Eeyore, that pessimistic, brooding donkey in A.A. Milne's *Winnie-the-Pooh* chronicles:

> *"Wish I could say yes, but I can't."*
> *"Thanks, but I'd rather stay an Eeyore."*
> Christopher Robin: *"It's your fault, Eeyore. You've never been to see any of us. You just stay here in this one corner of the Forest waiting for the others to come to you. Why don't you go to THEM sometimes?"*

Despair can arrest your will, motivation, and determination, as all three of the quotes illustrate. Gloom begets more gloom until you become mired in hopelessness until one day it starts feeling like the hopelessness is you.

I'll address the "doomer" perspective in more depth in chapter 10, but it's important to identify it here as a psychological barrier. Succumbing to despair is a zero-sum game; it deflates your energy, and squashes your sense of purpose and meaning in life. It can feel like a quick knee-jerk emotional response that overcomes you, or it can gradually eclipse you until one day you wake up under a blanket of depression.

Often, despair can cause a negative cognitive-behavioral feedback loop that runs through your head, locking you into a fixed mindset. For example:

> Thought: *We are all gonna die. We screwed everything up. What's the point of even trying?*
> Feeling: Small, helpless, afraid, weak, smothered, heavy, stuck.
> Behaviors: Less socializing, difficulty sleeping or oversleeping, irritability, low energy, reduced appetite.
> Thought: *Life sucks. This is hopeless. I give up.*

In short, while it's important to feel your feels and make space for them, there's a difference between *stewing in* them and *moving through* them (yes, the prepositions are important here). The goal is the latter. If you succumb to despair and hopelessness, it will eat you alive. When you feel crushed, sucked dry, depressed, and hopeless, it zaps you of your energy and motivation and unmoors you.

We'll discuss how to work with your emotions in the next chapter. In the meantime, if you're struggling with despair, try leaning on your friends, family, professionals, or community members to help you find your legs and regain strength, and reconnect with hope, create meaning, and a sense of purpose. Social support softens rigid defense mechanisms and gives you a boost of added resilience so you work through difficult painful feelings and experiences rather than flee.

Magical Thinking

"We can geo-engineer our way out of this thing."

Magical thinking, in the realm of climate, is thinking that someone else can do something that will suddenly improve things or undo the damage. It is an anti-scientific response, however, and stems from a human need to grasp for hope despite the odds. Denialism underscores magical thinking.

Why Parents Bury Their Heads in the Sand—and How We Can Look Up

(Hope, in itself, is a powerful agent of change, and a topic that I will more thoroughly address in the last chapter of this book.)

As I write this section, a recent article in *The Guardian* points to magical thinking by the aviation industry as they make a push to produce sustainable fuel as a potential solution to global warming.[6] While it is clear how this could be a game-changer in reducing a major source of greenhouse gas emissions, critics refer to this as both magical thinking and greenwashing because focusing on sustainable fuel production would steer important resources away from more immediate decarbonization priorities in an attempt to maintain business as usual. Time, as we all know, is of the essence.

Behavioral Contagion
"If others aren't changing their lifestyles, why should I?"

At the end of the day, despite American devotion to individualism and to "let freedom ring," we are herdlike in our behaviors. We copy each other's fashions, follow influencer advice on social media, and repeat stuff we hear just to be cool, liked, or to fit in. In less benign examples, we also join cults and carry out amoral transgressions when orders are given or succumb to peer pressure. Indeed, behavioral contagion is well documented, and it can be used for good or bad. So, if other parents are burying their heads in the sand, it creates an easy pathway to do the same.

On the flip side, however, behavioral contagion offers a remarkable path forward. Ideas, trends, and behaviors can positively ripple through communities. For a personal example, I'll tell you about a company, Ridwell, that collects unused items from your doorstep and recycles and repurposes different household items. There was an announcement at my kids' school about a collection drive, then a post on a neighborhood group chat, and the next thing I knew, Ridwell had exploded up and down our neighborhood. (Their trademark boxes are everywhere.) Many families are sharing the costs of the service to make it more accessible. Very cool to witness this (and to have our single-use plastics turned into Trex Company's composite decking)!

This is but an overview of some of the countless obstacles that prevent parents from talking, feeling, and acting. Worth noting is that both despair and magical thinking are types of all-or-nothing cognitive distortions rooted in distress. Thinking in such absolutes mires us in psychological sludge or obliviousness, neither of which is a healthy option for action. In the last chapter, I'll provide some strategies for moving out of rigid thinking. In the meantime, we'll take a stroll down the Yellow Brick Road of Climate Behavioral Change.

Diving into the Emotional-Behavioral Disconnect, Eyes Wide Open

When it comes to the climate emergency, disconnect between what we think and feel, and how we behave and act, is everywhere. An internal tug of war certainly explains poll discrepancies. A 2022 survey by the Yale Program on Climate Change Communication reveals that although adults are concerned about climate change (64 percent), there is very little discussion about it with family or friends (63 percent). Furthermore, 27 percent say they try not to think about global warming, and 15 percent try to avoid information about it.[7]

Since joining the online Climate Aware Therapist Directory, I've received a number of calls specific to climate anxiety. During my phone consultations, I hear and sense ambivalence in different ways. For example, a woman contacted me a couple of times, rescheduled consultations a couple of times, and finally vented to me while driving through rush hour traffic, offloading a twenty-minute verbal stream of existential angst, ultimately questioning if she really needs support. A participant in my "Parenting Through Climate Emotions" workshop was caught off guard by his anger and grief that surfaced about this topic. At home with his young child, he lamented that he'd avoided thinking about climate change for three years, especially because ecology was his field of study.

Why Parents Bury Their Heads in the Sand—and How We Can Look Up

Then there's a colleague's client who frequently voices distress about climate change but soldiers on with a jet-setter lifestyle, refusing to connect the dots. My neighbor, who excels at wildfire disaster preparedness but says "I don't talk about that stuff." I see all of these as expressions of disconnect between internal feelings and external actions, behaviors, and communication.

Parents care very much about ensuring a healthy future for future generations but often find it hard to sustain long-term engagement due to life. Anyone who grasps the enormity of what's happening is going to be worried, but worrying is not a comfortable place to be. We might take a few steps forward in acknowledging what is happening but then feel triggered or hopeless and retreat back into our previous ways. Or we might teeter on the "shoulds": I *should* get involved for the sake of my children, but I just don't have the time or energy right now. There might be all sorts of elaborate rationalizations or procrastinations: "Well, at least I'm donating money to Greenpeace" or "When my child is older, then I'll do more."

The results of a 2023 study by Hewlett-Packard suggest that parents are feeling increasingly anxious about climate change, and as a result are starting to align their lifestyles.[8] The study spanned parents' perspectives on climate change internationally, surveying five thousand parents from five different countries. Newsflash: a whopping 91 percent of parents are "concerned" about the climate crisis! The study determined that the majority of parents were reshaping their lives, careers, and purchasing habits based on their climate anxiety. That means that the emotional-behavioral disconnect is already shifting for some parents. But we have a long way to go, and many more parents to reach to amp up the Parent Club!

As you begin to take a closer look at your own disconnect, keep in mind that by disconnect, I mean feeling stuck, helpless, numb, and overwhelmed. These feelings manifest in a push-pull energy, leading to all manner of personal contradictions. While this disconnect can be dangerously insidious, with recognition it is also rife with possibility. Let's seize the opportunity here to examine it at close range, and learn to befriend it, so that we can move through it.

PARENT PAUSE

In this physical exercise, you'll build off the cognitive dissonance work that you did earlier around beliefs, obstacles, and actions.

Stand facing a wall, placing your hands on the wall an arm's length away. Call your beliefs to mind or even say them aloud, conjuring the energy that comes through believing in them. Push the wall away with your hands as you send that strong energy from your body. Hold that for fifteen seconds or so, whatever feels good for you.

Take a moment to reset and shake out your arms.

Next, call your obstacles to mind or even say them aloud. Place your hands against the wall again, and feel the hard wall pushing back against you as an obstacle. Feel that energy coming toward you, blocking you. Hold that for fifteen seconds or so, whatever feels good for you.

Last, put these two simulations together by pushing your hands against the wall while imagining your beliefs energizing you but encountering those obstacles. As you feel into this push-pull, imagine your commitment to actions as carrying you through the wall toward your goals.

Reflection questions:
- How did your internal landscape shift as you felt into these opposing forces?
- Were there any emotions or sensations stirred up?
- Were there any takeaways from this exercise?

Why Parents Bury Their Heads in the Sand—and How We Can Look Up

- Take some time to journal or share with an accountability partner. If you feel dysregulated, please visit Appendix C.

This exercise is intended to bring a felt sense, or bodily experience, to this idea of stuckness. If we feel it in this way, it might be easier to work with in our minds.

Introducing . . . the Yellow Brick Road of Climate Behavioral Change

Many who are privileged and/or satisfied in their lives do not wish to change things, relinquish their comfortable lifestyles, adapt to less, or give anything up. Call it what you will: desire, attachment, complacency . . . but it can be a compelling barrier to climate engagement. Our consumeristic habits and choices—such as eating meat, flying in airplanes, relying on single-use plastics (e.g., to-go cups and containers), and pushing the "Buy now with 1-click" button on Amazon—are addictive. And we know that although our unhealthy addictions are part of the problem, our carbon footprints are nothing compared to how much pollution companies are producing. But we still need to start by looking at our own complicity.

To increase your own awareness, grab a sheet of paper and write down a "Lifestyle Addiction Inventory" or a list of your unhealthy-for-the-MHW habits. This will help you to establish a rough eco-baseline. With addiction, it helps to be concrete and honest about frequency of use/behavior/type and so on. Before you can set goals or make strides, you have to know what you are working with! So take a moment to make your version of the list. And remind yourself that there's no judgment . . . just you, Parent Club member, showing up and trying your best to step up.

After you've identified your own sources of addiction, let's get acquainted with some ways to overcome our habits. Used by addiction counselors, the Stages of Change model is a visual tool for assessing a person's readiness to change. This tool is powerful because it creates a visual reference for both client and therapist to refer to and talk about, one that normalizes the rockiness of changing habits and addictions, and allows for conversations about addiction, barriers, relapses, and goals, for example.

In my work in climate psychology, again and again I find myself returning to the concept of addiction because so many folks are passive, avoidant, stuck, ambivalent, mired in habit, blissfully ignorant, and complacent in their lifestyles. (Phew!) But I've given this tool a creative makeover specific to climate behavioral change. First, let's review the original Stages of Change model and its six key stages in order to grasp its flow. It's typically a circle with relapse as an offshoot. Here's an example of a person's thoughts as they move through the Stages of Change:

Pre-contemplation: *"I don't have a problem to fix."* No recognition or intention to change; denial.

Contemplation: *"I might have a problem, but I'm not sure I'm ready to deal with it."* Exploring a possible change with hesitation; ambivalence.

Preparation: *"I'm going to try to change."* Shoring up motivation and support as you prepare to try.

Action: *"I'm walking the walk!"* Taking concrete steps and action toward your goal.

Maintenance: *"I must stay the course . . . I've come this far!"* Finding ways to sustain engagement and change, strategizing as needed.

Termination: *"I feel confident . . . There's no going back."* Old behaviors no longer have an appeal; stability.

Relapse: *"Oh nooo . . . I messed up big time. I'm awful."* Resuming previous behaviors, perhaps after triggering event.

It's important to shout out here that relapse is not failure! It just means that you've temporarily found your way off the wagon. It's okay, and it's often part of the process. No need to fear relapse or judge yourself too harshly. In the words of the late-1990s pop band Chumbawumba, "I get knocked down but I get up again. . . . " The Stages of Change model shows how it's possible for people to change by conceptualizing and talking about what that might look like.

And now, sit back, my Parent Club friends, and I'll take you for a stroll down the Yellow Brick Road of Climate Behavioral Change. The classic 1939 movie *The Wizard of Oz* is a tale of quest, longing, hope, love, growth, and overcoming fear and other obstacles. Please click your heels three times, and join me.

Okay, now for some explanation about this journey. As you read each description, consider where you might be right now.

Inertia. Like the Tin Man at the start of the Yellow Brick Road, you're frozen, unmovable, rusty, voiceless, and helpless. Defense mechanisms like the ones mentioned earlier in the chapter are afoot. Status quo is hard to break out of.

Teetering. Remember the scene when the Cowardly Lion, Scarecrow, and Tin Man are outside of the Wicked Witch's castle watching the guards marching below? Cowardly Lion is chosen to approach the castle guards to help rescue Dorothy. He puts on a brave face and musters his courage, before saying: "There's only one thing I want you guys to do." Scarecrow and Tin Man reply: "What's that?" The Cowardly Lion says: "Talk me out of it," and tries to run. You might find yourself grasping for excuses not to step up. Ambivalence and disavowal threaten to undercut your motivation, so you must be resolute in your decision.

Prepping. Before Dorothy sets out to run away from home, she packs a bag, puts her dog Toto into a picnic basket, and prepares for her adventure—even though she doesn't know exactly where she's headed. You might start initiating mental pep talks and sharing plans with others to help you prepare mentally to make changes.

Raising Anti-Doomers

Why Parents Bury Their Heads in the Sand—and How We Can Look Up

Engaging. Dorothy begins to walk tentatively down the Yellow Brick Road with Toto, one brick at a time, as villagers cheer her on. As she journeys on, she meets fellow travelers who accompany her, gathering strength, courage, and support. This is actually doing the work of individual climate behavioral change and, yes, you're on a journey with many other folks despite how personal it might feel.

Maintenance. Despite facing adversity (the Wicked Witch), setbacks (they are not allowed to meet the Wizard of Oz until they kill the witch), looming fear, and personal triggers (Scarecrow is set on fire, Cowardly Lion walks through the forest at night), Dorothy and friends do not give up. They stay the course, and support one another. That's what we all must do.

Lifelong Commitment. Dorothy wakes up in her bed in her home in post-tornado Kansas, her Auntie Em tending to her. She has journeyed, dreamed, and connected meaningful dots somewhere deep in her unconscious ("over the rainbow"), and this prompts her to see things with greater clarity, to deepen meaning and values in her life, and to love and feel grateful for those around her. She has integrated life lessons, dreams, values, and commitments with conviction, which will be with her for the rest of her life. When we face our "shadow sides," and find greater honesty and authenticity within ourselves, then we can truly commit to the greater good.

A reminder about relapse: At any point you might stumble, and that's okay. Step aside, and take a breather. When you're ready to regroup, you will. Our lives are multidimensional, and our timing differs. Most importantly, find your way back to the Yellow Brick Road.

One reason I like this lens for considering climate behavioral change is because it centers around the idea of *quest*. Although typically personal, quests often lead to bigger impacts or revelations, rippling out, affecting other people's lives. Can we, as parents, embark on a shared climate quest, for the sake of our kids?

Motivational Interviewing: A Tool for Managing Disconnect

Sticking with the notion that addiction counseling offers a path for supporting climate behavioral change, let's get acquainted with another addiction counseling technique: *motivational interviewing* (MI). MI is really just a compassionate and catalyzing conversation that helps elicit behavioral change and awaken motivation. MI can help us deconstruct deeply entrenched lifestyles, habits, and autopilot systems that naturally avoid thinking about and prioritizing big looming catastrophes, while at the same time strengthening inner motivation and commitment to change. MI is an intervention we can make in our own lives—with or without another person. When so much is out of our control, it's empowering and a relief to take this kind of action. Keep in mind that although MI is traditionally used by a counselor or therapist when working with a client, it can be adapted here and practiced. In the exercise at the end of the chapter, you'll have an opportunity either to do this with a partner or to try it out on your own.

Six Core Tenets of Motivational Interviewing

1. Listen compassionately
2. Ask open-ended questions (ones that move beyond yes/no answers)
3. Reflect back what you hear; summarize
4. Do not problem-solve unless asked to or until you have received permission
5. Validate ambivalence, while exploring discrepancies between feelings, thoughts, and behaviors
6. Roll with any resistance (do not push back or argue or try to win)

Now, let's see what this looks like in practice with this sample conversation:

YOUR FRIEND: "The wildfires in Australia are really disturbing. I've been feeling hopeless about the future—it's hard to focus on anything."

YOU: "I hear you. It's so devastating. . . . Sounds like it's affecting you to the core."

Why Parents Bury Their Heads in the Sand—and How We Can Look Up

YOUR FRIEND: "Yep. And I feel guilty, too. All this is going on while I'm debating between a latte or a cappuccino."

YOU: "It sounds like you're feeling some guilt. What do you think might help?"

YOUR FRIEND: "Well, I think it's really time for me to step up and do something."

YOU: "Like what kinds of things?"

YOUR FRIEND: "Maybe I could attend the local chapter of 350.org or something."

YOU: "Sounds like you are ready to take action. Good for you. I might join you."

In addition to MI, here are some other ways to move toward greater action and engagement on the climate front. As always, these tools are valuable for behavioral changes around most issues.

1. **Find a climate buddy or accountability partner.** Choose somebody you feel comfortable being vulnerable with. Mutually commit to stay engaged, no matter how small it may feel. Hold compassion for each other and remember that it's okay to stumble. Your climate buddy might be a good person to practice MI with. Be sure to check in regularly.

2. **Set up a climate date.** Set up a weekly or biweekly "climate date" with yourself or a climate buddy—as little as thirty minutes is okay. This can just be a space to process your climate emotions, or it can be attending a group or event, thinking through a lifestyle change and making a plan. Perhaps you might walk your dog with a friend once a week and talk climate or listen to a climate podcast. You do you.

3. **Move your body.** The magnitude of our climate emergency weighs heavily on our nervous system. The fastest way to kick-start your parasympathetic nervous system is by focusing on the body. Some ideas: taking a warm bath, snuggling in a blanket, walking, biking, yoga, deep breathing, dancing, martial arts, or exercise.

4. **Deconstruct a fat wad of overwhelm.** Sometimes mental gymnastics is required. Some therapists use the term "partialization," which means breaking something big into smaller parts so that it feels less daunting. For example, buying an electric car for the first time and getting used to it can feel like a lot to deal with. So perhaps you make a list of smaller steps that move toward it. And you take them one step at a time:

1. Watch a video about how EVs work.
2. Watch a video or find an article that compares buying versus leasing.
3. Compare available models and talk to someone you know who already has one about the pros and cons of an EV. (It's fair game to talk to someone random too when you're out and about.)
4. Test drive a friend or family member's car, or go to a showroom for a test drive.
5. Discuss financial options with a dealer.

How does that feel to you now? Try these steps with any big project or task.

5. **Realigning your values.** From time to time, most of us could use a values reset. If you haven't yet completed a values checklist from chapter 3, now is a good time. Or, you could try this angle instead:

1. Make a list of your values (www.motivationalinterviewing.org has a printout).
2. Rank your values in order according to what feels feasible for you.
3. Ask yourself or a partner, or write in your journal: How am I living according to—or in opposition to—my values? What shifts can I make so that I'm living in greater alignment with my values?
4. Be sure to pat yourself on the back for each change that you follow through on.

A PARENT CLUB STORY

A friend in Colorado has been struggling for years with a sense of panic and heaviness about the climate emergency. For a long time, she was stuck in a cycle of reading alarming climate headlines before bed, not sleeping well, and waking with an emotional hangover. Although able to acknowledge her emotions and articulate them to me, she felt overwhelmed by her family and work responsibilities and could not find the energy to move into Action on the Stages of Change wheel. As a New Year's resolution, she set the intention to finally do something. She took what felt like a scary leap and planned a community presentation at a local nature center. Her PowerPoint presentation was about climate anxiety and was open to the public, with time afterward for audience members to share. Participants expressed feeling relieved to have a space to talk about their concerns about climate change and wildfires. Despite initial insecurities about public speaking, my friend discovered that converting her emotions into action felt hugely relieving and energizing! I will get into this more in chapter 9, but my friend harnessed her climate emotions, took vulnerable action and forged a climate identity using her skillset.

Embrace the Hypocrisy—and Get On with It!

Imagine that I'm facing you, speaking to you in a soft, hypnotic voice: "You are all right just as you are . . . I hear you trying really hard to make a change . . . You're showing up and that's what matters . . . Sure, sometimes you'll say one thing and do another, but that's okay . . . Sometimes you'll act like the child and your child will act like the adult, and that's okay. Try to be gentle with yourself and others."

My point here is to shake you out of autopilot into imperfect action. Yes, taking action or even paying consistent attention might feel like a leap. And, yes, you might fall into a hot messy heap, and others might see you lying there. That can be part of the process. But leave the edited, airbrushed, and smooth sailing perfectionism for Hollywood screens. This is life, and it's raw and real.

I often joke with my friends that I feel like a walking hypocrite. One minute, I'm supporting a client with emotional regulation techniques so they can decrease their reactivity around their kids, and the next minute, I'm losing my sweet marbles because my child drew on the coffee table at our Airbnb rental.

As you prepare to make changes in your life and bump up against inevitable barriers, remember that social signaling—the behaviors that you model—is the most significant influence on others' behavior. What you do can ripple through the Parent Club, down the street, in the grocery aisle, or overseas. The Creative Exercise below will get you moving and grooving along the climate behavioral change track by combining elements of the Yellow Brick Road and MI.

CREATIVE EXERCISE

DIGGING INTO YOUR CLIMATE DISCONNECT

Materials:
- Journal, writng paper, whiteboard, or audio recorder
- Markers, gel pens, or dry erase markers (at least two different colors)

Why Parents Bury Their Heads in the Sand—and How We Can Look Up

Prompt:
Let's put some of these concepts into motion. Write out an imaginary dialogue between you and a counselor or friend, or even two different parts of yourself. (For example, your "moral self" and your "whiny entitled self.") You are welcome to dictate this aloud as an imaginary conversation into a recorder. If you have a climate buddy or partner, you can take turns in roles as speaker and listener, or even role-play as a therapist for some real fun. If you are doing the exercise on your own in a written format, you can select colors that correspond with the two different voices (e.g., red = therapist, blue = you).
Note: This exercise is meant to explore any "stuckness," disconnect, defensiveness, and ambivalence (among other things) around climate engagement to see what emerges. Feel free to consider which defense mechanisms you might be employing, or any helpful MI or Yellow Brick Road concepts, for example. If you are finding it difficult to get started, try using one of these prompts:
- I'd like to stop...
- I feel sad about...
- What keeps me up at night is...
- As a parent, my values are...

Process:
As you write or talk, notice—without judgment—any feelings, thoughts, and sensations arising in your body.

Product:
When you're done, take a moment to reflect, journal, or go outside and be present with your thoughts and feelings. Consider sharing/reflecting with others.

Reflection Questions
- What was the dynamic like between the two voices? Friendly? Hostile? Something else?
- What emotions came up for you?
- Did you employ any MI concepts as the therapist/listener?
- Did you learn anything new about yourself and where you are on the Yellow Brick Road?

PART 2

Kids

FIVE

Emotions "R" Us

*"There's a huge ball of hot rage in my chest that
erupts when the conditions are right."*

–Parent Club member

"I have a stomachache, but I don't know why."

–An eight-year-old child

While Part 2 of the book is geared toward kids, I will still be talking to the Parent Club first and foremost. Remember: We must put on our own oxygen masks in order to be present with our kids, and help them with their own coping strategies. To that point, as we forge into some deeper emotional processing, it will be wise for parents to do some of this work first *before* inviting kids to participate. What I recommend is that you read through chapters 5 and 6 and try out some of the exercises before involving your kids. That way, you will feel more prepared for how to respond to tough moments or questions with your kids when they come up.

One important distinction in Part 2 is that Parent Pauses are now Family Pauses, an invitation for parents and kids to do exercises together. Many of these activities are great for kids, from elementary school age to teenagers. Just keep in mind that you know your child best, and what might be appropriate versus too much or over their head. If any activity feels too advanced, I welcome you to simplify the language, scale back complexity, or hone in on one goal (e.g., to build empathy for animals or to draw/paint a recurring nightmare). You can also skip exercises that do not feel particularly well suited for your kids.

Climate distress is not always obvious or easily distinguished from other types of stressors. When I talk to other child therapists who are not focused on climate distress, they describe climate as one of many existential stressors adding to a general sense of unease and worry that simmers at the back of kids minds today much of the time.

As has been said already, climate distress is a normal and rational response to climate change. Emotions range from anger to anxiety, to grief to helplessness, to hopelessness to guilt, and so on. Often kids will experience emotions through physical symptoms like stomachaches, headaches, low energy, nausea, or insomnia—all clues of nervous system hypo- or hyperarousal. If your child is experiencing a new physical complaint that is uncharacteristic, it's worth examining any emotion present in a safe and respectful way.

If your child seems hyper-focused on climate or another issue, and they begin to experience symptoms like panic attacks, obsessive thoughts, sleep disturbances, depressed mood, loss of interest in daily activities, or thoughts of suicide or self-harm—or any notable change in mood or behavior that interferes with their daily functioning—seek out professional help. One of my hopes in writing this book is that readers will respond sooner rather than later to their child's climate distress (or even preemptively), intercepting before it spirals.

Emotions "R" Us

All emotions serve a purpose. They tell us who we are, what we care about, and where we have been. And, yes, emotions are also raw, and down and dirty. They are housed in the limbic system, that part of the brain that regulates emotions, memories, and behaviors, and that sits lower down in the brain than the prefrontal cortex. So when we dive into them, we're experiencing a literal dropping down. And we're typically dropping even further down into our bodies, since our body is scanning our environment for threats 24/7 and etching its perceptions into our nervous system with many updates.

The vagus nerve—the longest nerve of the autonomic nervous system—runs from the brain stem to the colon carrying signals between your brain, heart, and digestive systems. Vagal nerves play important roles in sensory and motor functions like heart rate, digestion, mood, and immune system response. This means that our bodies are constantly orienting to our external environment, and that our nervous systems are responding accordingly, prompting different moods, behaviors, and actions. Indeed, we store a good amount of emotional residue in our tissues, muscles, ligaments, shoulders, stomach, jaw, and other body parts through muscle tension, stomachaches, headaches, temperature shifts, heart rate change, and other physiological changes. That's why a massage feels so darn good at the end of the day, and why you might sometimes cry unexpectedly during one.

The physiological changes that accompany specific emotions are sometimes called "embodied emotions." In fact, studies have shown a cross-cultural component to how specific emotions activate specific body parts (e.g., anxiety is often experienced in the chest or gut). Some mind-body modalities focus on ways to release pent-up stress or trauma in the body, such as somatic experiencing, brainspotting, polyvagal exercises, and trauma-release exercises (TRE).

"Composting" and "metabolizing" are popular terms to describe deep emotional processing. I like these terms because they offer cyclical- and nature-based explanations of emotions. Emotions smolder together in a hot smelly heap, ripening and releasing signals to rouse you awake: Your emotions arise, and you choose to pay attention to some of them and retain some in memory, but you let go of others. Emotions are part of the greater cycle of life; animals from elephants to birds to fish experience emotions. In our quest toward reconnection in this book within the MHW, this is important to remember. Emotions are a part of being alive and are shared among various species.

To take this one degree further, philosopher Bayo Akomolafe offers important perspective on emotions and how they are embedded into our larger landscapes and systems, germinating in the space that exists, as he puts it, "between." In his words: "Emotion is not ours. It's not a brain phenomenon. It's a territorial phenomenon and it enlists bodies."[1] This is an important angle worth exploring because there's so much individualistic emotional processing these days that we can easily lose sight of the embedded nature of an emotion's origins. Moreover, that emotions are not as unique as we are inclined to think—that they arise in context and are often shared by others.

In this chapter, you and your family will learn how to harness emotions into action. I'll take you through a series of exercises to help you and your kids dive a bit deeper into your climate emotions with guidance and support. These exercises can be applied to any tough real-world issue. Please take PCPs as needed. If you encounter resistance or negativity, be curious and ask yourself where you think it might be coming from and why. We'll start out with a family-friendly exercise here to get you connected to your feeling state in your bodies. Less talk, more feeling, yes?

PARENT PAUSE

Land Sea Air Scan

Find a quiet place for this exercise. You may wish to dim the lights, set up yoga mats, face a wall, or do this in a quiet place outside if it feels safe.
Before we begin, think of an aspect of the natural world—an animal, plant or tree, a place that is peaceful to you (e.g., sunshine, a cloud, a beach). Take a few minutes to imagine yourself being there and connecting with it. Give this "friend" or place a nod or wave, and know that you'll be coming back soon.
Pretend you are an explorer of all that is happening in your body right now. Like a scientist, you will track and chart your discoveries without judgment. Whatever you find is just fine. Try to focus only on yourself and not on anything else.
Start by checking in with your body and see what you notice. You can wiggle or touch any body parts, or find movement to help you connect. What's happening with your skin . . . your bones . . . your arms and legs . . . your heart . . . your stomach? Are you hot or cold? Achy? Heavy? Tight? Light as a feather? Tingly? Grounded?
Now start to shift your awareness to your five senses. What do you notice looking around? What sounds do you hear? What smells? What textures do you touch? Is there anything that you taste? Find something pleasurable and enjoy that sensation. Take a couple of minutes to hang out here on your "land," your body.
Next, imagine that you are walking over to a safe and contained body of water. There is no possibility of any danger; this is a safe place, a refuge. Picture yourself entering the water in a way that feels natural. Do you wade in slowly,

jump in, dive, or take a boat? Picture yourself bathing in or touching the water. You are open, curious, and safe. Notice what the water is like: calm? refreshing? choppy? scary? powerful? Take a couple of minutes to explore this water. What feelings do you notice? Go ahead and name these, while remembering that you are safe where you are.

Finally, imagine riding up, up into the air in a hot air balloon. Let the experience unfold knowing that you are safe from any harm. You will tour the skies and then return safely to Earth to be greeted by those you love and trust. What's the weather like? Is it windy or calm? Do you have any worries or fears? Any thoughts or dreams? Spend a few minutes in your balloon allowing your thoughts to drift by while remaining curious. Then imagine descending down, down toward Earth . . . to be welcomed by those you love or trust.

As you come back to land, briefly revisit where you've been: land . . . sea . . . air. Breathe in the land (your body), the sea (your emotions), and the air (your thoughts) one by one, grateful for this exercise. Now, take a deep breath of gratitude and briefly revisit your nature friend or place that you connected to at the start of this exercise.

Take some time to write, journal, or create art and then share about your experience. Hold each person's experience with tenderness and curiosity.

What I like about this visualization is that it allows kids and grown-ups to get curious (but not judgmental) about different aspects of their mental health: emotions, thoughts, and body sensations. It's a way to understand that these different components are related: emotions are affected by thoughts, your body cues your emotions, your emotions reverberate in your body, and so on. Just like Earth's elements, there is a deep relationship between these aspects of being human.

For kids and grown-ups alike, managing emotions is essential for recalibrating mental health. If we cling to them or identify with them for too long, we become stuck. If we remember that they are fleeting, if we can, in a sense, surf them, then we allow them to run their course and keep a larger perspective. Emotional bypass—the act of quickly replacing uncomfortable feelings with positive ones—reflexively intellectualizes or ignores unpleasant emotions. That rejection perpetuates not only our suffering, but prevents us from taking necessary action from a heart-centered place. Emotions have incredible potential to energize and motivate. To that end, it's time for a brief lesson—a non-boring way to explore your climate emotions, I promise!

Demystifying Climate and Eco-Emotions

Okay, so you may have noticed that there are a lot of confusing terms splashed around headlines these days regarding mental health and climate change. In case you're scratching your head or trying to find a way to put your experience into words, here's a reference rundown. Although the words here focus on climate and eco-emotions, we can and should extrapolate, because many of these terms apply to issues at large. For example, pre-traumatic stress might be experienced by a child anticipating their return to school following a mass shooting incident. Dread might be experienced by someone who has continued contact with a racist or misogynistic coworker. A person suffering from long-term COVID-19 might experience disenfranchised grief.

- Ambiguous loss: Unclear and unresolved loss, including the loss of life's possibilities
- Biophilia: An innate human drive to seek connection with nature
- Climate anxiety: Often used synonymously for eco-anxiety, but focused more narrowly on climate breakdown and catastrophes
- Climate distress: An umbrella term that includes a range of climate-related emotions such as anger, grief, guilt, anxiety, dread, and trauma
- Climate dread: Anticipatory fear, worry, and grief prior to an event that we have little control over
- Climate emotions: Any and all emotions related to climate change (positive or negative)
- Climate empathy: A less stigmatizing reframe of climate anxiety as a prosocial response to environmental and social injustices

- Cognitive dissonance: A feeling of unease, helplessness, or distress when one's morals or values don't match up with their behaviors or their societal context
- Disenfranchised grief: The sense that others do not care or recognize your grief, and there isn't a place for it in society
- Eco-anxiety: Persistent worry about the changing environment or climate
- Eco-guilt: Guilt stemming from a sense of complicity in the climate crisis, or a sense of failure to stop it; also guilt associated with lifestyle choices
- Ecological grief: Both the tangible and intangible losses a person experiences because of the ecological crisis
- Moral injury: Injury to one's moral conscience stemming from a transgression, whether it be one's own or another person or group
- Pre-traumatic stress disorder: Anticipatory anxiety related to recurring climate or environmental events
- Post-traumatic stress disorder: Anxiety-related symptoms that develop after experiencing or witnessing a traumatic event
- Solastalgia: Grief associated with a personal experience of environmental change or loss
- Soliphilia: A feeling of protectiveness, love, and respect toward the planet
- Toxic stress: Chronic activation of the stress response system
- Vicarious trauma: An emotional response to learning about or witnessing the trauma of others

My colleagues at Climate Psychiatry Alliance, Climate Psychology Alliance UK, and Climate Psychology Alliance North America have crafted a free online resource, Ecopsychepedia.org, which provides clear explanations of climate and mental health terminology. If you're curious, check it out! Whatever your climate emotions are, they are appropriate. In the case of experiencing unwanted, maladaptive behaviors or emotions (or extreme symptoms stemming from very rational anxiety), it's critical to seek professional help.

For example, my colleague was treating a ten-year-old experiencing panic attacks and insomnia because she felt anxious about breathing; she was aware of how her exhaling put more carbon dioxide into the atmosphere. The general consensus among climate-aware therapists is that climate distress is a normal and rational response to the climate emergency and it should not be reflexively pathologized. Even as you and your family dig into your climate emotions, it's critical to resource yourselves. The toolkit discussed in chapter 6 highlights many ways to ground, metabolize, regulate, and manage emotions (if you could use some additional support right now).

> **WHEN TO SEEK PROFESSIONAL HELP**
>
> **If your climate distress is interfering with your ability to function: you can't sleep, feel deeply hopeless, you're crying a lot, isolating, experiencing panic attacks, nightmares, suicidal thoughts, loss of appetite, and so on, then it's a good idea to seek professional help.**

There are plenty of diverse resources for mental health support in the appendices. If you're experiencing moderate to severe distress, then individual therapy is a good idea. If your distress feels mild or manageable, then group therapy offers a good option. Some of my clients prefer to start with individual therapy and from there join a group to see how it goes, and expand their communities. Other folks are drawn into a group, eager to make connections and hear other perspectives, and sometimes feel therapy will offer additional support for material that comes up in group. Kids can do well with a safe place to just share and talk about their fears and worries. Offering a voluntary school or after-school group with a focus on a topic such as "big world feelings" or "friendships for future" can be helpful for kids. For those kids experiencing a higher level of distress or symptoms, individual therapy is more appropriate.

Dissecting Your Climate Emotions Pellet

Did you dissect an owl pellet when you were in school? It ain't pretty, but yes, I'm referring to the indigestible nugget that an owl regurgitates. It's actually quite fascinating because you never know what you're going to find in there—small rodent bones, teeth, fur, and so forth. Owls have efficient digestive systems for metabolizing small mammals, breaking down the nutrients that they need before upchucking the scraps. Yes, I'm grossing you out, but it's keeping you interested, is it not?

The Climate Emotions Pellet is a kid-friendly concept that I've designed to show the complex nature of our climate emotions, and how these might be festering inside of us. Moreover, just as with any science project, we're reminded to approach with curiosity, nonjudgment, and objectivity.

Just like the owl, we can't keep the pellet inside of us forever or we'll experience a chronic stomachache. You might approach this exercise by drawing or journaling what's in *your* Climate Emotions Pellet. Or, another

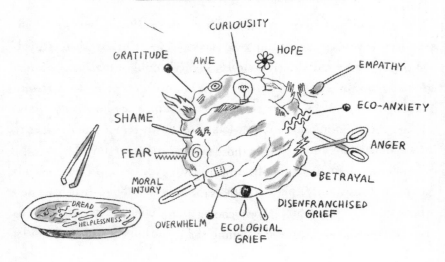

Emotions "R" Us

kid-friendly take is using modeling clay, Play-Doh, Model Magic, elements found in nature (mud, water, leaves, flowers), spices, or glitter. You can swirl colors together, use markers to add color, or create 3D mixed media pellets. There's no right or wrong way in how you choose to approach this. You could create the pellet without much thought and then dissect, or you could build the pellet using distinct colors or materials that represent different emotions.

Imagine delicately dissecting your own Climate Emotions Pellet with picks and tweezers. As you mine the pellet, you spread out your valuable contraband in an array to discover what's really there. Be curious and be sure to give yourself plenty of time and space to be with the emotions that turn up. Examine them, see them—and do so with gentle acceptance, not judgment. What might you be feeling, and what are you ready to let go of? If there's one emotion that resonates the most—as counterintuitive as it might seem—that is likely the path forward when you feel ready for it and adequately resourced.

FAMILY PAUSE

Take a moment to create a 2D or 3D version of your Climate Emotion Pellet with whatever materials beckon to you. Consider sharing these as a family or with a partner or group.

As you might have noticed, there are plenty of pleasurable emotions in the pellet. We can't forget that those are in there too! Not only that, but remember that we need to keep turning toward those emotions more than ever when we are confronting unpleasant emotions. In her podcast, Buddhist meditation teacher and psychologist Tara Brach says: "We cannot open to the sorrows without touching the joys." In other words, we must (as she sometimes puts it) "gladden the heart" because in doing so we create the space and ability to sit with darker emotions.

Now, truth be told, all of this emotional composting can seem privileged and self-indulgent. Compared to what those living in the most impacted and devastated communities are experiencing, emotional processing far removed from acute crisis can seem like a grand waste of time, even naive or farcical. But please stay with me here. In order to move toward commitment and action in a healthy, nonreactive, and sustainable way, you must first make this pit stop.

Many climate scientists agree that human emotions and behavior—not science—are the sticking point when it comes to taking action. The best way to enact change is to meet folks where they are at, and that means in real emotions they are experiencing or in the psychological scaffolding they've built up around them. Emotions offer a gateway into action, a way to move past complacency and despair. What are the risks of not processing emotions? Unprocessed fear can lead to aggression or isolation. Unprocessed anger can lead to hatred and violence. And so on.

In an interview with Jia Tolentino for *The New Yorker*, my colleague Leslie Davenport said that she pushes her clients to aim for a "middle ground of sustainable distress . . . [where they] become more comfortable in uncertainty and remain present and active in the midst of fear and grief."[2] This is really the difficult task that we are called to do for the sake of our kids—to push ourselves to a healthy threshold where we can stay engaged without burning out. It reminds me of how parents are charged with finding their emotional edges as soon as they launch into the crazy business of parenthood. Sharpening our edges, while staying connected to love and tenderness, can be viewed as another version of vulnerability as strength.

In that same article, Davenport points to the Catholic Serenity Prayer as a helpful guide: "God, grant me the serenity to accept the things I cannot change, the courage to change the things I can, and the wisdom to know the difference." There's a sweet spot of acceptance, vulnerability, honest limitations, and commitment that emerges if we can stay the course with our emotions.

Emotions "R" Us

CLIMATE EMOTIONS WHEEL

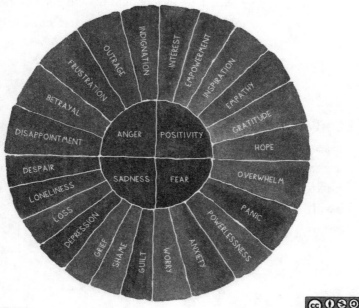

WWW.NICOLEKELNER.COM
WWW.CLIMATEMENTALHEALTH.NET
CLIMATE EMOTIONS WHEEL © 2024

ART ADAPTED BY NICOLE KELNER
FROM CLIMATE MENTAL HEALTH NETWORK

Identifying and Attuning to Your Emotional Body

In this day and age, there's a potential to get a bit heady and cerebral, even with our emotions. To counter this, this book is saturated with somatic content, so that we keep coming back into our bodies rather than spinning the wheels of our brain abstractly. The Climate Emotions Wheel is a visual developed by the Climate Mental Health Network (based on the research of Panu Pihkala at the University of Helsinki). We're going to look at this and use it as a jumping off point for connecting to our emotional bodies.

As you can see, the wheel is pretty straightforward in helping you to name and recognize your feelings. This wheel is also a great general reference for your family to explore any Micro- or Macro-level emotions. Why not keep it out on the coffee table for end-of-the-day discussions, or while watching the news with your family? Once you familiarize yourself with it, please glance over the basic somatic sensations words list in the box that follows. I've paired sensations words with the emotions wheel so that you can start to integrate your mind-body in a more holistic way.

> **SOMATIC SENSATIONS WORDS**
>
> **1. Starting Point**
> pleasant
> unpleasant
> neutral
> weird
> numb
> frozen
> tight
> tingly
> dizzy
> achy
> blocked
>
> **2. Temperature**
> hot
> cold
> warm
> cool
> neutral
>
> **3. Energy/Vibe**
> alive
> bubbly
> jittery
> pressure
> spacious
> heavy
> calm

If I had to name one key strategy that the majority of my clients avoid or resist but which ultimately supports their deeper emotional processing, it would be slowing down and sitting with bodily sensation. It's common for this total attunement to body to prompt unexpected tears. Most everyone I know spends a lot of time running around, chasing one thing and then the next, and resists slowing down. When they do, inevitably their emotions catch up with them. Those tears come with a profound bodily release—a

Emotions "R" Us

purging or catharsis. And guess what? They need to come out. Often they do so with a sense of deep-felt relief. Have you ever noticed how you feel better sometimes after crying? Or that you might laugh after a good cry?

This invitation to be present and give total permission to be with your experience is simple yet transformational. The pairing of your bodily sensations with articulating your emotions forges mind-body attunement and promotes healing and integration. From this place you can more safely investigate your responses to all that is happening in the world.

This Family Pause that follows is an opportunity to reckon more deeply with your sensations and emotions. I've included it in this chapter as a possible family exercise, but I leave it up to your discretion. The wording might be suitable for some teens, but for younger kids I'd recommend exploring sensations and emotions using simpler language and a mixture of positive and negative statements (e.g., the Earth is getting hotter because of greenhouse gases, Texas is the state that makes the most wind power). Or, I might have younger kids draw a simple outline of their body and use colors to represent different emotions and color where they feel them in their body when they think about an issue or topic (e.g., "anger in fists," "sadness in chest").

FAMILY PAUSE

Climate Emotions Layered Mandala

Grab a piece of wax paper and cut it into roughly five 8" sheets. Grab oil pastels and trace off approximately 6" circles onto each sheet. Have your Somatic Sensations Words and the Climate Emotions Wheel nearby for reference (you can always photocopy it). Find a comfortable seated position and take a few deep, clearing breaths. It can help to close your eyes or soften your gaze to help turn inward to your body's experience. When you feel ready, read or listen to each statement aloud, and then fill in each mandala using colors, shapes and lines. As you notice sensations in your body, write these down on the page or just say them out loud.

- Parents love and protect their children.

***

- As there are more floods, droughts, wildfires, hurricanes, and heat waves, more people will be affected – and many will have to move from their homelands.

***

- More than forty-four thousand species are now threatened with extinction.

***

- In 2024, deforestation (or cutting down trees) in Brazil's Amazon rainforest dropped by 30 percent.

***

- There are endangered species success stories including the Giant Panda, the Bald Eagle, the Steller Sea Lion, the American Alligator, and the Southwestern Black Rhino.

When you're done, take a few deep breaths and wiggle or move your body, however feels good to you. Stack the pages on top of each other in whatever order looks right to you. Staple them or glue them together.

Reflection questions:
- What sensations and emotions came up for you in this exercise?
- Share page by page.
- Look at it as a whole.
- Which pages or emotions stand out the most? least?
- Perhaps give it a title.

Please take a beat after this exercise and check in with yourself. What might be helpful at this juncture? Perhaps practicing self-care, a PCP, journaling, sharing with a friend or climate buddy, visiting the resource list in Appendix J?

This exercise offers an integrative experience for processing climate emotions. Too often we stay in the high reaches of our prefrontal cortex, spinning out on statistics or headlines without acknowledging the deeper emotions residing in our bodies. Hopefully, this exercise has helped you and your family to reorient so that you are more aware of your feelings. The reset that is in order is for your mind to pay attention to your body and trust it (rather than other way round). The concept of an emotions mandala can be used again and again in your life any time you're curious about your emotions or feelings. For more fun with mandalas, check out the Climate Emotions Mandala Project that was spearheaded by integrative art therapist Mor Keshet and journalist Anya Kamenetz.

Good Grief, Grief Is Everywhere

One of the deepest, rawest, and most profound life-altering emotional experiences is grief. Grief is a complex muddle of many different emotions and sensations: sadness, anger, confusion, numbness, heaviness, fear, love, guilt, shame, heartbreak, and hope. Grief underscores our response to the Macro-level climate crisis, mass shootings, war, the pandemic, poverty, and violence. Yet, as discussed in chapter 1, it's also part of our everyday personal lives in the form of the loss of a loved one, a breakup or argument, miscarriage, ailing health or a new disability, aging, and so forth.

Grief is rife with dialectic emotions and sensations; it has been part of the human experience since the dawn of time. And yet how we express it—or don't—has been shaped by our patriarchal-dominant culture. In today's America, grief is often seen as a sign of weakness, thereby suppressed, scorned, and bulldozed. One of the most universal, frequent, tender, and vulnerable emotions gets hardly any airtime. This is another way that cognitive dissonance plays out in our culture today.

So many of us are walking around deeply grieving past or present wounds. It's hard to find the time to grieve—or be acknowledged in grief—because it's not sufficiently legitimatized. Or you could say we just don't feel as comfortable talking about it in the same way we do the Super

Bowl or the Oscars. It's sad when I hear in my office how folks don't want to be a buzzkill with their grief, so they isolate themselves or don't bring it up. It's also sad, though totally understandable, that friends, partners, or family members cannot bear to support others through their grief, likely due to their own unresolved grief or trauma.

Many sit with this uncomfortable emotion silent and alone; it can feel as though you are being eaten up from within. Or we grow up not knowing what to say or how to relate to someone who is grieving. We fail to show up in supportive ways even as we long to. It can feel agonizing to bear another's deep suffering. While in the throes of grief, we might feel unseen by our society, a feeling known as disenfranchised grief. This denial of grief is harmful for our bodies, minds, spirits, and communities.

For kids, this happens early on in an adult tendency to encourage kids to gloss over and move on from emotions (e.g., "please don't cry"). Yet grief is present in the early years, too: from parent separation or divorce, to loss of a pet or loved one, to health issues, or even the birth of a sibling. Allowing kids space and permission to be sad or angry and fully grieve will help them to integrate the harder parts of life as well as understand that they are part of what it means to be alive.

Let's do some of our own grief work now! For many of us, this is long overdue.

Grab some basic materials to draw your Grief Graph. This concept comes from *The Grief Recovery Handbook* by John James and Russell Friedman. Kids can do this too, and you can explain that it's about sad times, happy times, and disappointments in your life.

Create a simple *x-y* axes, and on the *x*-axis place your age lifespan. On the *y*-axis, you will be charting the significant highs and lows in your life (e.g., high = a graduation, new job, or new relationship; low = death of a loved one, a car accident, or starting a new school). Place your dots at different ages to mark the highs and lows. Then connect the dots in a line.

Keep in mind that any time you have experienced an unmet hope or expectation, grief can be present. For example, the tough decision not to have a child, a failed marriage, or an illness that prevented you from doing something.

Reflection questions:
- What have been your highest highs? Lowest lows?
- What can you learn about your life from this graph?
- How do you relate to your grief? To other's grief?

Please journal or share with a buddy as appropriate. If you feel triggered, please visit Appendix C.

To sit with grief—or many of these emotions, for that matter—is hard. But it's a required process. To paraphrase Tara Brach, true courage is not rooted in overriding fear or pretending—it's rooted in honest contact with our vulnerability.

The Pandemic Afterglow

One of the biggest, most under-regarded, pandemic "gotchas" is what I call "revenge of the screens." Before the pandemic, experts warned parents about the negative effects of too much screen usage on children's brains and bodies, particularly on their mental health. During the pandemic, with parents stressed to the gills, everyone gave them a free pass. Fair enough. But afterward, there wasn't a righting reflex of resetting norms and expectations around screen use. In a sense, residual grief and trauma from the pandemic, among other factors, has steered parents toward digital acceptance.

Screen use—from gaming to smartphones to social media—continues to soar, even as researchers like Jonathan Haidt have pointed out evidence of increased anxiety, depression, self-harm, and suicide attempts as social media use on phones accelerated between 2011 and 2015.[3] Our kids are in desperate need of a postpandemic digital reset; screens are one of the reasons why their emotional health is tanking. Just listen to my tween client who said that she begs her parents to help her manage her screen use more because it feels overwhelming to her, and the depressed teenager who told me matter-of-factly, "Everybody my age is addicted to their phone."

I could write many pages on this subject, but the reason I bring it up in this chapter is because screens create emotional barriers—between your kids and their emotions, kids and their families, kids and nature, kids and exercise, and so on. To do the work in this chapter—in this book, for that matter—requires discipline around screens. When your child feels sad, it means that they don't self-medicate by scrolling social media. When your child is angry, it means they don't reach for their phone to post something. Yes, social connection is important, but so is learning to feel, respond to, and regulate your own emotions.

In my practice, when kids and teens are stressed or depressed, I tell them or their parents to get their bodies moving and get more fresh air. Sometimes I sense a parental eye roll, but they usually realize that they're not doing enough physically. And I almost always hear that it was comforting or energizing for them to spend time outdoors—whether it was walking the dog, sitting in a yard doing homework, taking a phone call on a walk, biking, or hiking. Too much screen use is making our lives feel more harried and less fulfilling! We can't underestimate its effects in the mental health equation.

As we wrap up this chapter, please remember that these emotional tools for exploration are available at any time. They are tools for navigating our uncertain future, expanding insight and capacity to tolerate our own distress, and for teaching our kids how to tolerate theirs.

A word about trauma: If you find that you're struggling with emotional expression and feel stuck, consider seeking professional help. Sometimes trauma can inhibit our full range of emotional expression.

CREATIVE EXERCISE

STAINED GLASS EMOTIONS

Materials:
- Black marker or Sharpie
- Crayons
- Electric warming tray (optional)
- Heavy paper
- Tape

Note: If you don't have a warming tray, then you can use oil pastels and watercolors instead (but I highly encourage you to get one if you can).

Setup:
Use marker or Sharpie to draw an outline for your stained glass. If using the warmer, turn it on to medium heat and tape a sheet of paper on top of it.

Prompt:
If using oil pastels and water colors, fill in stained glass with oil pastels as described. Feel free to experiment by layering watercolors on top for a wax-resist effect. Use intuitive colors, shapes and lines to show what it feels like to be your age (or to show all of the feelings inside your heart).

Process:
As you create, notice without judgment any feelings, thoughts, and sensations arising in your body.

Product:
When you're done, tape your work up on the wall, and look it over with childlike curiosity, and without judgment. Spend some time with the reflection questions that follow, and journal or share your creation with others.

Reflection questions:
- Let's practice what's called the "TTAQ" in art therapy shorthand:
  T= title
  T= theme
  A= affect or mood
  Q= what question does this piece ask?

SIX

A Twenty-First Century Anti-Doomer Toolkit for Families

"Sometimes I just need to do something for myself, ya know?"

–Parent Club member

As we gear up to face today's issues and challenges—climate chaos, gun violence, racism, war, the pandemic—we must sharpen and expand our existing toolkits. And we must help kids do the same. Doing so will increase resilience in our communities, and it will help our kids develop anti-doomer mindsets that will serve them through life.

It's important to distinguish between two different approaches to self-care: the first being as a way to tune out, avoid, or escape life, and the second being to bolster resilience and mitigate burnout. This chapter focuses on the latter. I firmly believe that regularly unplugging, taking mindful breaks, time-outs, and retreats is part of the long-term solution for sustained advocacy, activism, and socio-emotional engagement.

Readying a Cross-Cultural Toolkit

The toolkit I offer in this chapter is an imperfect attempt to synthesize an array of psychological, emotional, physical, physiological, and spiritual tools from various cultures and traditions. A toolkit is inevitably values-based, and therefore ethnocentric; after all, tools and solutions are borne out of one's own cultural beliefs and traditions. It's my goal to create an inclusive, integrative toolkit that honors various cultural traditions related to centering practices, resilience, and body-mind wellness. After all, in this era of converging crises, our cross-cultural wisdom offers collective sustenance and enormous potential to repair a fractured, ailing global community.

Please bear in mind that seeking, relating to, finding inspiration, or adopting customs from other cultural traditions must be done with great care and respect, deep humility, and a gentle sense of inquiry (ideally with a representative guide or mentor). When a person intentionally or unintentionally fails to acknowledge an element of culture by a member of another culture or acts in a disrespectful or exploitative manner—be it a ritual, tradition, or practice—then they cross the line from cultural appreciation into cultural appropriation.

Keep in mind that this chapter could be a book in and of itself, so I'm sharing just a few ideas and concepts that might anchor, inspire, build confidence, and support your family in forging ahead. If a particular practice resonates, then trust that direction! There is obviously much missing from this section; let this be an invitation to add in your own elements and sources of inspiration. Next, I'll introduce the Existential Body Map (EBM), as a way to organize and conceptualize all of these rad tools.

Why a body map? Well, the simple answer is because we will be calling on our bodies in their entirety—and not just on our overindulged prefrontal cortexes—to elicit our full evolutionary aptitude for resilience. The EBM is designed from head-to-toe, concretizing a range of activities, approaches, and exercises to try out to bolster your resilience now and into

A Twenty-First Century Anti-Doomer Toolkit for Families

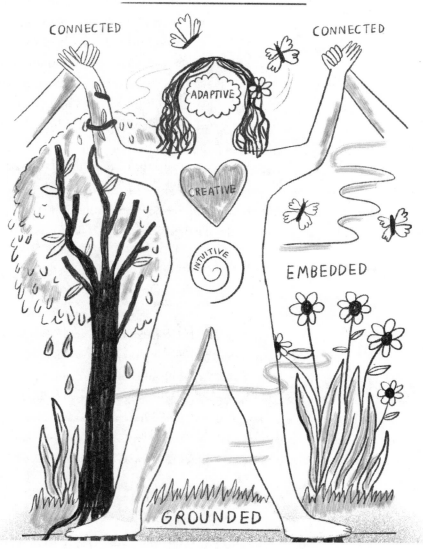

the future. There are six categories, each of which is associated with an aspect of the body:

- Adaptive (head)
- Creative (heart)
- Intuitive (belly)
- Grounded (feet)
- Connected (hands)
- Embedded (human-environment)

I'll begin each category with a brief family-friendly exercise to zoom in and bring awareness to that particular aspect of the EBM. As you read through this sizable section, you may wish to jot down notes for your own personalized EBM. You can sketch a body shape and jot notes around the various body parts so that you remember those nuggets that speak most to you.

Adaptive (Head)

Find a comfortable position, sitting or standing, indoors or outdoors. Take a few rounds of cleansing breaths. Start to notice your breath, your body, and what's around you. Gently pet your forehead and hairline with the palm of your hand, bringing softness and compassion for all of the thinking that it does. Gently massage or pat your facial features (nose, ears, eyes, lips) and your neck. You can also soften your gaze by lowering your eyelids slightly. You are letting this part of your body part know that you are present, curious, and open to exploring whatever comes up.

Parents: As you read this section, keep toggling back to this awareness, using touch as a way to periodically reconnect.

A Twenty-First Century Anti-Doomer Toolkit for Families

How many times have you heard someone say, "I'm in my head a lot?" Well, as a therapist, I can tell you that I hear this nearly every week.

Isn't it strange how a tiny portion of our brains—the prefrontal cortex—dominates our days? Our cognitive abilities like complex reasoning and logic are what make us uniquely human. Yet, concentrating so much of our energy in that realm does us a disservice if our thinking brain overshadows our sensory experience. Many of my colleagues, myself included, have reoriented toward body-centric modalities in recent years, such as sensorimotor, somatic experiencing, polyvagal techniques, tapping, and Hakomi.

Here are some practices you can do to recalibrate your cognitions so that you experience a greater sense of flexibility, spaciousness, and open-mindedness—all important elements for the road ahead.

Plugging into the present moment. For many of us, living out of sync with the present moment—and often in a near constant state of anxious or future mind—has become an unfortunate default mode. Psychologist and meditation teacher Tara Brach calls this the "trance of forgetfulness."[1] Think about a time when you drove from one side of town to the other hardly aware of how you got there because you were lost in your thoughts. If we are disconnected from ourselves, how can we authentically connect with others? Moreover, how can we remember the climate emergency or war that's happening in a country far away?

While writing this book, it's become crystal clear that "modern autopilot" mode might be the death of us (literally). By modern autopilot (MA) I mean that default setting by which we do things automatically, habitually, sans awareness, and instinctively conform because, hey, that's easiest. But it's also an existential pitfall or trap. MA threatens our collective ability to act and undermines our mental well-being all in one fell swoop. To break out of our habits, whether it's nail biting, excessive drinking, or thoughtless consumption, it takes awareness, motivation, and follow through. The good news is that simple actions can instantly disrupt unhealthy patterns.

As a benign example of MA, at different points in my life I've woken up and drank a cup of coffee and re-caffeinated throughout the day trying to stay alert. In the last few years, however, I've questioned and adjusted this habit. Rather than wake up each morning and make a cup of coffee, I now *feel* into the level of caffeine that I need; in other words, I consider what my body feels like it's asking for, how long and stressful my work day may be, how much energy I have in that moment. Our needs are in flux, after all, and tuning into our bodies' sensation and nonverbal cues can help to right our balance.

Note: This is not a Family Pause but rather a Parent Pause.

- **What's your version of MA?**
- **Spend a moment reflecting upon any habits, behaviors, or patterns that might not be serving you, your kids, or a particular social issue. How might you gain awareness in the moment to challenge MA?**

When I'm working with a client who is struggling to break a habit, I often advise them to find moments of pause during their day, both planned and unplanned. Examples of planned pauses are scheduling a walk during a lunch break or taking a few minutes to breathe deeply before entering your front door at the end of a work day.

These moments are rich with opportunity. And trust me, you will fail again and again, as we all do. But starting to build pause into daily routines and reduce your reactivity in stressful situations will puncture your habits and restore a new degree of choice and freedom. The space between stimulus and response is where choice lies. In the moment, it can feel like there is no choice—that our reaction comes instantly. But what if you visualize or draw out a continuum with a pause or buffer between an identified stimulus and response? This will visually reset a visceral experience that you can conjure up in your mind the next time. For the parent about to yell, there can be three deep breaths or even five push-ups before acting. For the child starting to feel angry or anxious, they can take a bathroom break or stretch to cool down. Brach calls these types of acts "the sacred pause." When we encounter stress, she says, we need to be able to pause to allow light to come through.

Whether firing up breath, reflecting on an affirmation, or emotionally regulating in another way, this will help you or your child to snap out of mindless MA whirlpools and reestablish mental flexibility. The simple act of breathing supercharges our bodies to act by oxygenating our minds to follow through on our intentions.

Bye, bye screens! Plugging into the present moment also requires *unplugging* our devices. The addictive nature of smartphones can wreak havoc on your mood, self-esteem, nervous system, and concentration span. Just think of the rings, the dings, the chimes, and alerts! Unplugging gives your whole being an opportunity to just simply be.

When we are free of mental distraction—and stop the epidemic of multitasking!—we can physically engage with our five senses, daydream, imagine, create something, and experience a sense of awe and wonder and vitality. For inspiration, I often look to animals: the sound of a cat licking its fur on a chair in the sunlight; a dog's excitable sniffing the air and ground when it first leaves the house; a duck shaking water droplets off its wings.

Now for the good news: There are plenty of ways to unplug! Many of my teenage clients are required by their parents to hand over their devices at a specific time each night. Some parents participate in this ritual as well (or at least feign it until bedtime). Others make a joint agreement to unplug from dinner to bedtime to be present with their families. I've been experimenting lately with one weekend day unplugged until noon. Usually, it's a day when everyone wakes up leisurely, and there's no rushing to get out the door. I don't check my phone until midday, and it feels—incredible!

When my older daughter gets off the iPad, she often blurts out: "I'm bored!" It's clear to me that what's really going on is a shift from passive "couch potato" mode to an active state requiring some energy and motivation. Usually, it takes a moment for her to reconnect to her body and her interests. Maybe she has a snack and we brainstorm ideas together. What holiday or birthday party is coming up that we can prepare for? What games, books, and toys are around her? Who's that playing at the park across the street? What projects has she started but not completed? Plugging back in (with some support) helps to revitalize her, returning color to her cheeks.

"Name it to tame it." This catch phrase was coined by interpersonal neurobiology pioneer Dr. Daniel Siegel to describe how parents can help kids to verbalize their experiences and thereby help them to regulate difficult emotions. This strategy also applies to adults. Getting granular with climate emotions makes them less overwhelming. Emotions have different shades, qualities, and textures that affect us consciously, unconsciously, physically, physiologically, and spiritually. Simply naming those shades, qualities, and textures is fundamental to taming them. By now, you've done some of this work using the Climate Emotions Wheel.

A simple yet powerful meditation that is a gateway to help your kid better understand their emotions, and for you to better understand yours, is Tara Brach's "RAIN" meditation. Here is its essence:

Recognize (what you are feeling)
Allow (rather than judge)
Investigate (why this feeling might be coming up)
Nurture (welcome this feeling as part of your experience)

This tool is really a twofer because it both plugs you into the present moment and guides you through sitting with and naming uncomfortable feelings.

The next time you or your child feels rattled, try listening to Brach's thirteen-minute audio or make your own way down the RAIN list. Try pairing this with deep breathing, movement, or sitting in nature so that you don't get coiled up in your mind. If RAIN is too difficult to grasp, then you can translate it into a kid-friendly poem, such as: Hello to my feelings/let them be just as they are/hmmm . . . why might they be here?/I'll make space for them now. And remember that the reason that we are labeling emotions like this—particularly those hard ones—is because facing them means that you will not be under their spell in the same way, that they will likely reduce in intensity, and they will not snowball in the same way as they do when they are repressed. For kids, making space for emotions teaches them early on that they are valid and deserve recognition—creating healthy coping patterns that can last a lifetime. Furthermore, kids will learn to be responsive rather than reactive. For more inspiration on this topic, check out the poem "The Guest House" by the thirteenth-century Sufi poet Rumi.

Challenge and replace "ANTs." Automatic negative thoughts (ANTs) are self-defeating scripts that replay in our heads, negatively affecting our moods and outlook. I've seen many, many clients over the years who feel paralyzed, hopeless, or perpetually insecure because their internal record player is stuck in a groove of negative self-talk or doomer talk.

Take, for example, this recurring thought: "The climate crisis is hopeless. Nothing will ever change, and the government isn't doing anything so we're all doomed." Say that aloud three times and see how you feel. Scared? Depressed? Heavy?

Now try tweaking that thought so that it still resonates as true, but with a greater sense of hope. For example: "The climate crisis is real, and it's scary. But I'm part of a growing group of people who care about others and about the Earth and we are all working hard to turn this around." How do you feel now? Replacing negative thoughts with more hope-infused positive reframes of a situation can help alleviate a chronic sense of doom, anxiety, or despair.

In my office, I challenge ANTs all the time. I often ask child clients to write down their ANTs on a whiteboard or on paper so they can objectively witness just how easily and frequently they assault themselves with this kind of talk. Seeing their own thoughts so concretely often prompts an eyebrow raise, nod or giggle of recognition, or an "ah, I see now!"

moment because it so clearly reveals our negativity bias and the incessant self-defeating chatter in many of our minds. Some of my clients will come up with names for this negative internal voice (e.g., negative Nelly, worry wart, inner critic). Doing this helps expose the voice for what it is—just a reflexive voice that's taken up residence in our mind! And it will then grow smaller, losing power over time.

Building off this, the next step is to imagine and write out a compassionate reframe for each ANT. Compassionate reframes are kinder, more hopeful and thoughtful evaluations of a situation. By flexing this voice, you tap into an indelible hope that resides in the human spirit, a wellspring of creativity. For many it can be hard to reframe an ANT, and it takes practice and repetition. It can help to summon up the voice of someone supportive, loving, or wise in your life. This really is a life practice that will grow easier as you retrain your brain and build new neural pathways. With kids and adults alike, it can be helpful to create puppets for the ANT voice and the compassionate voice and name them and even practice reframing with them (brown paper bag puppets work great . . . add on as much as you want with googly eyes, yarn hair, and other adornments).

A few points for getting started:

- Start to notice ANTs in real time and call them out: ANT! Beware!
- Keep a running written log throughout the day
- You can practice compassionate reframes in the moment, or come back to them later when it feels less charged
- If you feel stuck, pretend you are doing this for a friend or a child (and notice how it's easier)
- Option: fold a page in two lengthwise and use different color markers for the two columns, ANTs and Compassionate Reframes (see Appendix A)
- Observe how you feel in your body and mind *after* reframing your thoughts
- Breathe this feeling in to help cement an energetic shift in yourself

A Twenty-First Century Anti-Doomer Toolkit for Families

Attitude makes a difference. Research has shown that we can rewire our brains, attitudes, behaviors, and beliefs. If you *believe* they can change or grow, then . . . you do! An oft-recited mantra in school classrooms is: "I can't do this . . . yet." A growth mindset is a belief in change or possibility—and sustains engagement, hope, and motivation. The idea of a growth mindset is taught early on in elementary schools to help foster a positive learning environment and illuminate a sense of potential that resides in all of us. The belief in growth and possibility is an important one for sustaining hope on the road ahead. On the other hand, if we bow to negative attitudes or doomerism, then we limit our potential.

A quick word on gratitude: Gratitude is an attitude toward life that will serve you well, especially if you can tap into it daily. Feeling into a grateful heart will set a positive stage for the day and soften whatever sorrows and wounds come your way. Consider a morning and/or evening practice, even for a couple of minutes.

To wrap up the Adaptive section, please take a moment for a Parent Pause or Family Pause. And speaking of choices, you can choose which of these you'd like to try out:

- Practice RAIN
- Start an ANT log with compassionate reframes (see Appendix B)
- Reflect on your attitude and how you might tweak it to take back a sense of agency and hope

Meanwhile, intense events will continue to play out on the Macro and Micro levels. There is nothing that we can do about that until we try. By checking our negative assumptions and our attitude, we restore power, choice, and agency in how we show up for our kids.

Creative (Heart)

Find a comfortable position, sitting or standing, indoors or outdoors. Take a few rounds of cleansing breaths. Start to notice your breath,

your body, and what's around you. Read or listen to the passage that follows.

Long ago, our ancestors communicated to one another through cave drawings and paintings (using minerals, berries, and animal blood), through gestures and grunts, and eventually through language. They played instruments, sang, chanted, drummed, danced, and performed rituals. Many cultures do this today. Raw creativity is deeply encoded in our human DNA. Our survival depends on it. Human beings are creative creators. Each one of us is a creative creator, no matter what we may think or others may say. Even more, our survival depends on it.

What thoughts, feelings, and sensations arise in hearing this? What's your response? Feel free to discuss.

Reclaim your creativity. Often we're discouraged in our younger years from our creative gifts by adults and peers. Often their criticism gets the best of us, and we abandon this part of ourselves. I frequently hear comments from adults like: "I'm not creative! I'm bad at art! I'm no artist! I can't draw." Researcher Brene Brown has referred to this as "art shaming," and unfortunately it has discouraged many of us from creative pursuits or creative modes of expression. But that stops right here, on this page of this book!

The therapeutic benefits of art-making are vast! To name a few: emotional processing, nonverbal communication, trauma healing, emotional regulation, meaning-making, insight building, and community bridging. Just the sheer act of kinesthetic engagement—such as rolling clay between your palms or painting with a brush laden with thick pigment—is pleasurable and triggers reward pathways in your brain. When you feel freedom to create something in a safe, nonjudgmental environment, it can bring you into a "flow" state of quiet focus, deep calm, and sensory immersion to the point that you are no longer aware of time. Have you ever experienced this?

A Twenty-First Century Anti-Doomer Toolkit for Families

Here are three ways to jumpstart your creativity with your kids.

1. *Five senses immersion.* Gather your kids and, if possible, some art materials or pens and paper or journals. Note: You can do this spontaneously without anything on hand as well. Sit outside amid birds, sky, trees, flowers, chirping birds, scuttling insects, and MHW for some *en plein aire* fun! Spend some time engaging your senses and sharing: What do you smell? Hear? Touch? Taste? See? Take time to enjoy tuning into each of your five senses and sharing with each other. Sometimes it's fun to see things from a different perspective (e.g., lying in the grass or climbing up a tree). Everyone can write, draw, paint, or build in response to their sensory experience. If you don't have any materials, gather found nature allies (e.g., branches, rocks, leaves) and create a personal or communal sculpture. Remind everyone to turn the dial down on their thoughts. Remind them to create or put marks or words onto paper in any way they feel like and to trust their intuitive process. If you notice any judgments arising, gently wave them along—they are not needed. When you're finished, sit quietly and wait for the group. When sharing, be curious and kind in how you listen to others and share your own. This fusion of nature bathing and creativity is a double dose of well-being. Note: If you created a sculpture, consider if it's an offering that is okay to leave as is or if it feels right to restore the area back to how it was. Be aware of your reciprocal relationship with the MHW (what are you taking, giving, or leaving behind?), and teach your kids to be mindful, too.
2. *Hop on the kid train.* Does your child want to create a Lego tower? Make a comic strip or flip book? Dance or play music with you? Sometimes parents unnecessarily clamp down on kid creativity because they are worried about the mess or don't think it's worthwhile

or convenient. But what if . . . you toss your parent agenda aside and join 'em instead? They will love you for it. Join your kids in coloring, drawing, Play-Doh, food art, collage, stamping, and so on. If there are some fun recyclables to use (e.g., cardboard, gift wrappings, corks or bottle caps, food scraps), even better!

3. *Say, what time of year is it?* Holidays, birthdays, and seasonal transitions are good markers for being creative. Mural-making is a fun, low-pressure group exercise. Our family hangs an Earth Day mural on our garage door, sometimes with painting materials, so others who walk by can add to a community mural. Seasonal art activities might include leaf rubbings, botanical prints, flower pressing, painting pine cones or rocks, using natural dyes (e.g., turmeric, berries, avocado pits, coffee), or pumpkin seed jewelry. You can roll out some butcher paper or a tablecloth and set up a space outside on a patio or inside, then gather the materials and get started! Sometimes it's helpful to create an example of the project either beforehand or in real time. If you share an example, be sure to tell your child that this is just your example and theirs will look different.

While paint and other materials are flying around, you might be surprised at how comfortable it is for your child to express themselves in this way. Art-making can feel less confrontational and vulnerable than straight up talking. Plus, everyone is more emotionally regulated as they pound, scratch, scribble, paint, or glue. Oftentimes words bubble up naturally alongside art-making. Talking can happen during or after art-making. Some of the benefits of art-making in a safe context are calming the nervous system; expressing unconscious feelings, desires, hopes, and fears; and connecting with others through a positive activity that builds trust and connection. Even if you're allergic to the idea of art therapy, my guess is that you will soon see how creativity can benefit your child, alleviate distress, and guide your family in solving hard problems!

A Twenty-First Century Anti-Doomer Toolkit for Families

Conceptualizing the mandala. "Mandala" is Sanskrit for "circle," and its design is typically a circular pattern radiating out from it center. In art therapy and Jungian psychology, mandalas are considered cross-cultural symbols of wholeness, wellness, and balance. During this time of social fragmentation, mandalas can be beacons of light. Viewing or making mandalas can help to rouse us out of collective disassociation, create order and integration, find meaning amid chaos, repair our sacred relationship with our MHW, and reestablish a sense of community and belonging.

In these tumultuous times, mandalas are powerful images that we can meditate on, create, color in, witness, walk inside (e.g., a labyrinth), or simply notice that they exist in the MHW. Just the act of creating or witnessing a mandala can help to heal ourselves from within, as well as relate to a sense of balance and wholeness that exists in the universe—even in its depleted state. For example, planet Earth is a mandala, as is the sun, a full moon, flowers, snowflakes, and spider webs. Can you think of some other examples?

The box "Get Your Mandala On!" offers some family-friendly ways to create or engage with a mandala.

GET YOUR MANDALA ON!

- *Spontaneous mandalas:* Wherever you are, you can use available resources to create mandalas, whether it be in the sand on a beach or playground, stones or flowers, chalk on concrete, seaweed grasses on the beach or fallen branches and leaves on a forest floor. Tibetan Buddhists offer much inspiration through their construction of intricate, time-consuming mandalas of colored sands with small tubes and funnels, which are eventually destroyed as a metaphor for life's impermanence!
- *Outdoor labyrinth walk:* Have you ever stumbled upon one of these maze-like circles in a nature area or park? At the center, people have left offerings, blessings, painted rocks, messages, or similar items. To try a labyrinth

walk, select a stone or other meaningful object, and stand at the outside entryway of the labyrinth. Holding the stone, walk slowly and mindfully into the center, contemplating this question: What's something that I'd like to let go of in my life right now? When you arrive in the center, place the stone on the ground, and take a few deep breaths, imagining a releasing. When you feel ready, walk out, slowly and mindfully, contemplating the question: What's something that I'd like to take with me (i.e., an intention)? It can just be a word or a phrase. Try to trust your intuitive process. Let the phrase wash over you as you walk out of the labyrinth and back into life.

- *Mandala "pie" response art:* For this group project, draw a large circle onto poster board, and divide and cut up into pie-like slices. Each person paints or draws artwork around a shared theme (e.g., a tragic community event, climate emotions). Collectively piece together the mandala on a board or paper. Hang up and view it in its totality.

Experimenting with new creative frontiers. Some other expressive creative modalities follow. Many of these are offered by specialized expressive arts therapists. The main takeaway here is to be willing and open to step out of your comfort zone and try new things. Zen Buddhists call this *zazen*, or beginner's mind.

Dance. Dancing is fun and cathartic, and there are many cultural ties to dance that elicit memories, traditions, identity, and social cohesion. Some studies suggest dancing is associated with improved memory, promotes social connection, and lowers risk of dementia. If you would like to shake your body in a safe space, here are a few ideas:

- Zumba
- Salsa (or Meringue, Bachata, Tango, etc.)
- Modern/Contemporary
- Line dancing
- Square dancing

A Twenty-First Century Anti-Doomer Toolkit for Families

- Five rhythms
- Ecstatic dance
- Tap dancing
- Hip-Hop
- Bhangra/Bollywood
- African dance
- Belly dancing
- Break dancing
- Roller skating
- Family or friend dance parties
- A wedding or other event
- Your local dance studio
- Privacy of your own home or room

Kids naturally gravitate toward shaking or wiggling their bodies to music. During toddler years, many parents will sign their kids up for a dance/movement class, and gawk at how cute their kiddo is. But then one day it's like everyone forgets that dance is more than an activity to excel at—it's also therapeutic. Dancing is a cheap and accessible form of fun that offers an emotional outlet, opportunities for social connection, and brings you closer to joy and awe.

Why should dancing be constrained to weddings, special occasions, dance classes, and school dances? Sure, you may have to play more Taylor Swift than you'd like to, but let your kids have fun DJing and grooving. I know many families who, during the pandemic (and beyond), hosted family dance parties. Amp it up with a disco light, or invite friends or family too. Play music in public spaces and invite open-air dance parties. (There's nothing more pleasant that stumbling upon a group of sweaty people dancing with huge smiles on their faces.) For me and other mom friends, dancing has been a profound form of release and tapping into awe. And—just to be clear—many of us have zero training as a dancer. But we do have a willingness to shake our bodies and laugh at our many mistakes in attempting to follow choreography. Two of my friends with intense careers—in public health and climate finance—have started teaching dance classes outside of their day jobs and they're positively gleeful.

Music. Both playing and listening to music are deeply healing. Yes, sound vibrations are a real thing and affect your mood. Again, playing an instrument, singing, humming, or whistling are body-based, patterned repetitive movements, which, as you know by now, help regulate your nervous system. One study found that playing Mozart reduced heart rate and blood pressure.[2] There's a primal association with safety and comfort in gentle rhythmic sound; babies in utero sense that rhythmic backdrop of mama's heartbeat.

Incorporating music into your life can be done quite easily. On the way to school in the morning, my kids understand that it's classical music time (or what they call "mama music"). But at home or on road trips, the kids play their music. You can spice up plenty of household chores with music, and it can be comforting and satisfying no matter the genre (okay, no promises on death metal). You can do it in community and experience the powerful grounding that comes from attending a concert, drumming in a circle, playing in a band, or singing in a choir.

For a fun family-friendly exercise that you can do outside, first form a circle. One person starts off with a sound that they repeat over and over again (i.e., beat box, hum, ocean sound, whistle). When the next person is ready, they chime in with their own sound, and then the next person, and so on. Once you are all going, enjoy the nervous-system-settling fun and vibrations. The first person will eventually stop, then the next person, and everyone else until the last person has stopped too.

A few other examples to try:

- Listening to music outside
- Sound baths
- Drumming circles
- Kirtan
- Singing in a choir

A Twenty-First Century Anti-Doomer Toolkit for Families

- Playing in a band or orchestra
- Listening to a musical
- Opera
- A capella
- Exploring different genres, depending on your mood

For many teens, music is life and an essential part of their coping. A favorite group art directive among teens is to design an album cover about your life and include a playlist. Teens love to explore and broadcast their identities. Ask your kids: are there any songs that are important to you (or for parents: any songs that have been important in your life)? Go ahead—talk among yourselves. Or play some of the songs and look up the lyrics. Sometimes kids don't want to talk about emotions, but they are willing to play a song that resonates with them and talk about the lyrics. Many kids are interested in writing songs and even performing them. Remember that music can be passive or it can be active, and it's versatile!

FAMILY PAUSE

What music or songs inspire you or your kids to emote or act?

Consider making playlists because they can be a tangible tool in your toolkit to come back to. Some ideas: a Climate Grief mix, a Joy mix, an Activist mix. If you are in a depressive spiral, then you could listen to tunes that match your mood, or you could try listening to tunes to help pull you up. If you know that you are blowing off the important stuff, like taking climate action, then it might be time to create space for your emotions, passions, and grief.

Storytelling/Writing/Theater. Whether spoken aloud or written down, the act of putting your feelings into words can feel both relieving and helpful for communicating with others. Hearing other people's stories offers inspiration and a sense of companionship and validation at a time when we could really use it. Finding courage to express your authentic voice is the first step.

Performance offers the possibility of catharsis, allowing emotions and energy to flow outward through speech, movement, music, song, dance, and dialogue. This is just as evident in your child's school play as it is in street theater protests. For some inspiration check out the Insure Our Future campaign by Mothers Rise Up.

Keep in mind the potential of drama therapy as well. Following 9/11, drama therapy was an intervention used to help school children in New York City cope with its traumatic aftermath. Spoken word has also been a positive outlet for young people to express their struggles through words and performance. The 2022 documentary *Our Words Collide* follows five high school seniors at the onset of the pandemic who experience poetry as a positive outlet in their lives.

In the adult space, there is much we can also garner here. Playwright Mimi Stokes recently offered a dramatic eco-psychology visualization for a group of climate-aware therapists (myself included). She prompted us to reflect on a hopeful outcome for the climate crisis. By tapping into creative expression and its imaginative possibility, she helped to conjure up in us "superabundant awe," a foil for tragic awe. Powerful meditations like these can help us to remain hopeful and circumvent doomerism. Joining a group can feel supportive and help hold you accountable in your storytelling in whatever form it takes.

A few years back, I helped organize and run a couple of climate-focused poetry workshops for therapists. These workshops created a safe container for emotional unpacking and expression of eco-grief as we heard from different pockets of the country, sentiments ranging from the dried up cricket fields in England to the noted absence of insects on car windshields on the East Coast to a young adult's yearning for a sense of safety amid wildfires in Colorado.

A Twenty-First Century Anti-Doomer Toolkit for Families

Yes, you can absolutely write or tell stories on your own. That said, there is much healing that can happen in doing it as part of a group, or family or community. Joining bodies and voices and hearts together is an immensely powerful action. I'll speak more about social connection later on, but it's a double-win to combine creativity with a social group. That's why protests, drumming circles, street theater, and similar activities are so powerful. Creativity—in whatever form—starts with your openness and curiosity and spirals out into community in contagious inspiration.

WAYS TO GET CREATIVE THROUGH STORIES, ACTING, AND WRITING

Tell your story . . . or just listen to others!
Climate Stories Project: climatestoriesproject.org
The Climate Journal Project
Kiss the Ground–Stories for Regeneration: kisstheground.com/storytelling
Writing children's books
Writing "cli-fi" (science fiction with a climate focus)
Self-publishing
Community theater
Street theater
Improv groups
Open mics
Poetry
Writing prompts
Writing circles
Podcasts
Social media post or blog
Puppetry

Intuitive (Gut)

> Find a comfortable position, sitting or standing, indoors or outdoors. Take a few rounds of cleansing breaths. Start to notice your breath, your body, and what's around you. Gently place one hand on your belly, and think of kindness, compassion, and curiosity spreading out from your palm across your belly. You are letting this part of your body know that you're present, curious, and open to exploring whatever comes up. What sensations are happening in your stomach right now?

Parents: As you read this section, keep toggling back to awareness in this region, using touch as a way to periodically reconnect.

Intuition is undervalued in Western culture as a way of knowing. Science and logic reign supreme. Parenting is a good example of this: rather than follow our own hunches and rhythms, many of us comb through books, blogs, and reels searching for the perfect answer to a particular issue (be it sleeping through the night, poor behavior, or sibling fights). Oftentimes the answers to our questions lie within ourselves. But in our data-obsessed culture, intuition is latent, underappreciated wisdom because it's so often hijacked by reason and logic. Just think of the last time you were pulled one way instinctively, but reasoned yourself out of it. While helpful to keep our impulses in check, sometimes we end up throwing out the baby with the bathwater and we disregard a fundamentally human way of knowing and experiencing.

Intuition is an instinctive feeling or impulse that lights the pathway forward. Intuition comes through bodily sensations, such as a heart flutter or tingling of the skin. Other times it comes as an idea mysteriously nudging you in a direction. What's important is to recognize that these messages are impulses arising from your deepest core to tell you something important. And it's time to listen now more than ever!

Note that on a physiological level, both intuition and the body's trauma response can sometimes feel similar, and therefore difficult to discern.

Both might cause you to feel tingly, hot and cold, or like you suddenly feel on high alert and your heart is pounding. If there's a negative memory or flashback associated, that's likely trauma. On the other hand, if there's a feeling of openness and curiosity that impels you in a direction, that's likely intuition.

Here are some tools to help you lean in to your intuition.

Call in your wise guide. Many child therapists use sand trays in their offices. Children are invited to select figures from a collection of different animals, mythical creatures, or archetypes (e.g., the unicorn, the warrior, the mother, the sea turtle, the superhero). A child will intuitively enact resonant scenes that tell, imagine, or reimagine a story in some way. The sand tray is powerful with adults as well. Sometimes the figure a client selects will turn out to be a wise guide, other times I might direct them to select a wise guide. A wise guide can be carried with you at all times, both literally and figuratively. And it's an exercise that your family can do regardless of age.

In some shamanic traditions and some therapeutic orientations, you may be directed to call in or visualize your own wise guide. This guide can be actual or imagined, human or nonhuman, gods/goddess, animal, an element of nature, a superhero or character, a spiritual figure, or an ancestor. I'm referring to a wise guide here in a very general sense—of calling on a figure or element to help you along this journey as a parent. Note: Shamanic journeying is a practice that involves entering an altered state of consciousness to connect with spirituality for healing, insight, and guidance. The term *shaman* is of colonial origin and is problematic in that it conflates distinctly diverse Indigenous forms of healing. New Age forms of shamanism often skew into appropriation. The Family Pause that follows is included as a way to explore spirituality and altered consciousness as a source of wisdom and resilience, but it is not linked to any specific cultural traditions. It is an exercise to connect with your wise guide as a touchstone of resilience that you can call in again and again.

Raising Anti-Doomers

FAMILY PAUSE

For kids and parents:

Help your child look through images in books and magazines of characters, avatars, and action figures, or draw them from *their* imagination, anything that is accessible. Have them pick a wise guide that is right at that time. What might they consider its powers or attributes. What might you learn from it?

For older teens and parents:

Take a moment to set the stage for this quiet, reflective activity inspired by shamanic journeying. Dim the lights, close the door, and create a comfy floor space to lie down or just lie down on a couch or bed if available. It can be helpful to listen to soft rhythmic music or beats (e.g., nature soundtracks, drumming, chanting, or instrumental music; see my website for links). You can select a soundtrack of five to ten minutes or set a timer. When you're comfortable, close your eyes and remind yourself to trust this process as you listen to the music and soak up the vibrations in your body. Wait and see what images arise in your mind's eye. Observe with curiosity, and follow to see if your wise guide is revealed. When you find your wise guide, be curious and see what they have to teach you or what journey they take you on. When the song ends or the timer goes off, do not rush back. Take your time to slowly transition back into the present, stretching, and blinking eyes open slowly. Perhaps journal or share.

Reflection questions:
- Who is your wise guide? What attributes do they have (e.g., stay calm under pressure, shape-shift)?

- How was this exercise for you? Were you able to fully participate or were there any barriers?
- If this exercise did not work for you, you can try a different way in to Wise Guide as discussed in a moment. Think: Who or what embodies the wisdom that you need right now?

Connecting with a wise guide can empower you through difficult times. If it feels too far out to do without some help, fair enough! This work can absolutely be done in spiritual circles or in some therapeutic traditions (EMDR is one example). Another way in is to simply look for an image, figurine, or photograph that speaks to you, art-making, breathwork, dreamwork, guided visualizations, hypnosis, memories, photographs, magazines, or SoulCollage. See what you gravitate toward.

Be sure to connect with your wise guide regularly: preserve it, photograph it, draw it, record an audio of yourself describing it, or anything that makes sense to you. When you feel lost or overwhelmed, you know where to turn. In that moment, picture them approaching you, acting to protect you, or imagine how they might act in your situation. Wherever you are, you always have your guide to turn to for answers.

Be aware of cognitive dissonance. As mentioned previously, cognitive dissonance is a felt sense of disconnect between your inner and outer landscapes. For example, we may feel one way inside, but behave contrary to this feeling to satisfy external conditions. This is an alert to your system that something doesn't feel right. As an example, some folks have shared with me that they love to travel, but as climate change accelerates, they agonize about flying on an airplane. If they keep up the same level of travel, the cognitive dissonance—experienced through intuitive body messaging—will only intensify and disrupt their inner peace. There is an element of cognitive dissonance that is rooted in intuition.

And there is most certainly cognitive dissonance in how we go about our lives despite the magnitude and urgency of our polycrisis. To help bridge this gap, we must trust our intuitive felt sense that things are not okay and heed that red flag. From there, we can start to align our inner thoughts/feelings with how we live/behave (such as capping our number of flights each year or purchasing carbon offsets).

One of my clients anguished over working at his business-as-usual-job as the climate emergency worsens. As he began to address his cognitive dissonance by exploring and moving toward greater congruence with his values, he experienced levity and relief even as he maintained engagement.

Body intelligence. As a new mom, I came across an interesting book while browsing in a local bookstore called *Body Intelligence: How to "Think" Outside Your Brain and Connect to Your Multi-Dimensional Self* by John Mayfield. The book introduced a Japanese concept called *hara* to me, which is a spiritual point of strength located in the lower abdomen that some Zen Buddhists and practitioners of martial arts believe is essential for holistic well-being. Hara unifies physical, psychological, and spiritual realms. Becoming more attuned to this region of my body has been a gateway for cultivating my resilience. There are various ways to connect with hara, including through breath, meditation, yoga, physical therapy, and dance.

The idea of body intelligence—of locating a place in your body that feels like a spiritual home that you can physically access—is helpful for the road ahead. Hara is one concept, but there are others. Ask yourself: Is there a place that lights up inside of you when you're happy and relaxed, that tingles, emanates, hums, or feels like a kernel of wisdom? For some people, it's their heart space, for others it's their belly or soles of the feet. Find the place in *your* body that feels strong and capable to hold you through tough times.

As life unfolds around you, the tool of body intelligence will continue to serve you if you stay attuned. Part of our collective work as a society is a willingness to expand beyond our own limited experiences, stretching our belief systems, and sometimes our cultural comfort zones. This means thinking outside traditional Western paradigms such as pharmaceuticals, conscious reasoning, or even talk therapy.

INTUITIVE MODALTIES

Here is a shortlist of intuitive modalities for kids and parents to explore.

Kids and parents:
Craniosacral therapy
Osteopathy
Naturopathy
Ayurvedic medicine
Body shaking practice (from qigong)
Martial arts
Drumming circles
Sound baths/vibrations
Massage

Parents only:
Reiki
Acupuncture
Traditional Chinese medicine
Psychedelic journeying
Shamanic journeying
Holotropic breathwork
Women's, men's, or nonbinary circles
Sweat lodges
Cold plunges
Natural hot springs

I'm not telling you to try anything that feels off or out of alignment or triggering. Only *you* can decide what is healthy experimentation on the road to building your intuitive toolkit. And the same goes for making parenting choices that feel right to you.

Laughter. Have you ever seen your kiddo running around and trip over something and fall down, but instead of crying or yelling they actually laugh at loud? And then everyone else laughs too? Laughter can cut through tough stuff. Laughter gives you that same rush of discharging intense energy as crying or yelling, yet it's actually pleasurable. Most kids love being silly and will look for any excuse or partner in crime. When stressed, it's something that parents can quickly forget about—but it's always a way to diffuse emotional charge or even just connect after a hard day.

In 2023, I attended a reading by activist Andrew Boyd for his book *I Want a Better Catastrophe: Navigating the Climate Crisis with Grief, Hope, and Gallows Humor.* At times, despite the topic of climate change, the audience laughed so much that you'd have thought he was doing a stand-up routine. There was a visceral sense of relief to be together and find humor despite the heaviness.

As studies indicate, laughter is a free resource that actually reduces physiological tension. In the short-term, laughter increases your intake of oxygen, stimulates your heart, lungs, muscles, and circulation. Laughter also increases endorphins released in your body and helps your muscles relax. In the long term, laughter improves your immune functioning by releasing neuropeptides, relieves pain, and improves mood. In our busy, action-packed world, we could all use a few more laughs each day. And, of course, it costs nothing and is readily accessible.

So go ahead! Break for laughter. Each day. Whether it be talking to a person who makes you laugh, listening or watching a TV show, funny reel, or stand-up clip, doing a little laughter yoga (which is a real thing!), or being silly dancing or clowning around with your kids. Try giving yourself permission to find the funnies in life—it will buoy you!

Play. Kids are the experts on how to play. In essence, it's their main activity in their early years, and it's what they know how to do. Think: digging in the sandbox, playgrounds, tag, make-believe, forts, and dress-up. That said, the adult world operates at such a different cultural tempo that it can squash child-centered play. Psychologist Jonathan Haidt has pointed out that as smartphones gained traction, we are living through a great "re-wiring" from play- to screen-based childhoods (roughly between the years 2010 to 2015). He points to evidence of a spike in youth mental health issues that coincides with access to smartphones.[3] For today's parents, who are under tremendous pressure amid the intensity of an all-consuming digital culture, one way in which they leverage some control is to resist the temptation to substitute screens for play. In addition, they can leverage some control by banding together with like-minded parents to protect a healthy play-based childhood, and not fall prey to techno-optimism or groupthink. Child psychologists, pediatricians, and child educators recognize that children have an innate urge to play, experiment, build, create, fantasize, imagine, and socialize, which is important for their positive growth and healthy development. Even as it may be inconvenient for parents at times to disappoint our kids, it's our job to nurture body-based, kinesthetic play both outside and indoors; in doing so, it will build resilience.

Now for the adults. Remember when, as a child, you lost yourself in your play? You might have built forts, played hide-and-seek or sports, explored the neighborhood, played dress-up, or rode bikes? Playing, like laughing, is a fundamental part of being human. In a workaholic culture, it's greatly endangered for adults.

But it's critical that adults do not abandon our ability and willingness to play and make play a priority. When we are faced with something as momentous as the climate crisis, we sometimes instinctively curl into a forlorn ball and shelve play for the sake of our focus and commitment. But balance is essential, and play offers much-needed physiological relief and lightness for our darkest hours and most demanding moments. Some grown-ups do things like dancing, trapeze arts, attending concerts, frolicking on the beach, personal getaways, take bubble baths, travel to new places, or go on retreats. Other grown-ups may jam in a band, play sports, knit, make Lego creations, attend formal social events, plan scavenger trips, paintball fights, and road trips, and reinvent themselves as they learn new hobbies.

FAMILY PAUSE

For parents:

What was one of your fave ways to play as a child?
What sparks your joy/interest/curiosity nowadays?
Are there any activities or events that you'd like to try out or return to?

For kids:

What are some of the ways that you like to play that don't involve screens?
Name one thing that you would like to try out, practice, or try to get better at.

Take moment to journal or share.
Perhaps, if you are feeling excited, set an intention to follow through.

Play is housed in the intuitive section of the book because it really is an innate human urge that is increasingly recognized by neuroscientists and psychologists as a source of energy, creativity, and motivation.

Grounded (Feet)

Find a comfortable position, sitting or standing, indoors or outdoors, barefoot optional. Take a few rounds of cleansing breaths. Start to notice your breath, your body, and what's around you. If standing, try pressing the soles of your feet down into the floor or earth below, and feel into that contact and gravitational pull. You can play around with this by tapping your toes or lifting your heels up and down, or by imagining grounding the four "corners" of your feet by gently rolling

them. Wherever you are, come into stillness and feel the sensations in your feet and legs. Imagine that part of your body growing roots that shoot far, far down into the earth like the roots of a tree, spreading out across the soil. Spend a few minutes noticing what it's like to have deep roots that support you.

As you read this section, keep toggling back to awareness in this region, grounding your feet as a way to periodically reconnect.

Since the days of Freudian psychoanalysis, talk therapy has dominated the therapy space. But in the 1970s, somatic and mindfulness-based approaches began to gain traction in the mental health field incorporating breathwork, meditation, and movement. In my view, and that of many others, mind-body healing cannot be split—and we still have a way to go since Descartesian dualism. To supercharge our resilience toolkits in this day and age asks us to reconnect with our bodies and heal what needs to be healed.

In recent years, studies have focused on our autonomic nervous system (that regulates heart rate, blood pressure, temperature, digestion, sexual arousal, etc.) and how it influences our thoughts, feelings, and behaviors. Psychologist Stephen Porges introduced polyvagal theory in 1994 and emphasized how the vagus nerve, which spans from the brain to the large intestine, is a central component of the parasympathetic nervous system (i.e., "rest and digest"). His theory explains how physiological states of arousal influence our daily lives from moment to moment, including our mental health. In other words, we shift between states of mobilization, immobilization, and relaxation. He and his team designed specific exercises to help reset the nervous system. Porges' colleague, psychologist Peter Levine, developed Somatic Experiencing in the 1970s, a body-oriented approach to healing trauma, which has been complemented by Porges' subsequent research.

As noted in Porges' research, the brain-gut axis is bidirectional, meaning that your stomach is primed to send information to the brain based on sensory input from the body (it's not just the brain dictating everything!). This explains the gut response phenomenon where your stomach seems to know before the rest of you catches up about how your body is interacting with the environment. Porges' work has invited mental health practitioners to focus on techniques for regulating the autonomic nervous system as well as helped to educate and normalize human stress and arousal states.

In recent years, there's been a greater emphasis placed on what British psychologist Kimberly Wilson calls "whole-body mental health," bringing nutrition, exercise, sleep, and lifestyle to the forefront of the mental health equation. These body basics are foundational for mental health. The adage "we are what we eat" has been proven in a sense through research showing that gut bacteria determines our mood. It's surely good news that nutritious food can positively influence our microbiome and improve our mental health. (Although there are equity dimensions in what food folks can afford and have access to.) In short, the health of our bodies is central to our mental health and contributes to our wellness.

Befriend your nervous system. Let's be honest: in a cerebral, screen-engaged world, we sometimes forget about what's below our chins. After a day on the computer, we stand up and our bodies feel stiff, achy. My husband, who is in back-to-back Zoom meetings all day, often forgets to eat lunch. It's practically an act of resistance to prioritize our own needs—or family's needs—during the workday.

Whether you acknowledge it or not, your body is communicating with you all of the time. Yes, right now, feel into it . . . notice any sensations, temperature shifts, rumbles, pulsations, aches, energy, or emotions? On an average day, we constrict and tighten and tense areas of our body, often unknowingly. Gaining body awareness and learning to respond accordingly soothes anxious nervous systems.

As parents, we have the agency to do this on our own—*and* to help our kids cultivate body-centered practices early on. Kids naturally gravitate toward shaking out their sillies, running, wrestling, dancing, playing tag, and so on, and this is a natural advantage. What kids are inclined to do—and need space and time to do—are exactly these patterned, repetitive movements that activate the parasympathetic nervous system, reduce cortisol levels, and promote healthy brain development. Staring at screens will hardly provide this opportunity, so it must be carved out in other ways.

I remember my younger daughter felt anxious about trying an after-school theater program, complaining that her stomach hurt, and she didn't want to participate. I quickly realized that our conversation was going in circles. She had a response or comeback for everything. So I distracted her by initiating a clapping and singing song and we clapped hands together and sang. It was like pushing a giant reset button. She smiled, then giggled, then shrieked, and begged for more. When the program started five minutes later, she took her place on line. Because she felt calmer in her body, she was able to try something new.

Credit: *The Zones of Regulation Digital Curriculum*, Copyright © 2025 Leah Kuypers. All rights reserved. Used with permission.

"Zones of Regulation" is a socio-emotional tool often taught in elementary school curricula, and it happens to be great building block for the family. Your child can probably explain the model to you. In it, feelings are grouped into color blocks representing different states of the nervous system: calm (green), slightly aroused (yellow), aroused (red), and numb (blue). The idea is to teach kids to notice shifts in their feelings and sensations and talk about them.

You can either print out the graphic or have your child create their own version by writing their feelings for each color bucket, as well as identifying triggers, activities, or moments when they experience each of these color zones. Parents can participate, too, and you can share together as a family. Other variations on this concept featured in Appendix C are the Window of Tolerance; the River of Life; and Owls, Watchdogs, and Possums.

Breathe. While breathing sounds like an easy thing to do, it's also not as simple as it seems. Often, kids in my office stare at me blankly when I ask them if they remember to breathe when they feel worried or anxious. "Breathe!" as a command is a paradox—and yet it's nonetheless mission critical.

We offered some breathing exercises earlier in the book, and it truly is trial and error for kids and adults to figure out what works and what is realistic to use in a public situation. During a stressful moment, your body tenses, your nervous system moves out of the Green zone, and your prefrontal cortex shuts down and reduces your decision-making capacities. As mentioned in chapter 3, your breath is your best friend, so use it! When you notice your breathing grow shallow, try firing up some deep belly breaths. Connecting to your breath by attempting to lengthen and slow down your inhalations and exhalations is your fast-track ticket toward emotional regulation. The same applies to kids. Here are a few other ways to practice grounding, when you notice your breathing change, while engaging your senses more fully.

TIPS 'N' TRICKS FOR EMOTIONAL REGULATION

In the moment:
- Bare feet in the soil or grass
- Singing, humming, whistling, growling
- Walking, swinging, jumping rope
- Coloring books
- Fidget toys
- Making a cup of herbal tea, inhale the scent, exhaling and blow the tea to cool it, then repeat
- Listen to music, dance, or shake your body

Got some time?
- Guided meditations using an app, setting a timer, playing soft music
- Gardening with your bare hands
- Calm cards (a kid-friendly card deck of mindfulness exercises)
- Swimming or paddling
- Playing an instrument
- Hand or foot massage with oil or lotion (Newsflash: We discovered that our kids *like* to give us massages! Said my youngest the first time it happened: "Parents usually take care of kids, but we are taking care of you guys. I like doing that.")

Breathing and grounding practices might be old news to you, but they are an art that we practice and sometimes move away from until we find our way back again. More than half of the battle of self-care is resetting the equilibrium dial on your nervous system.

Yoga. Love it or hate it, yoga is one of the best grounding tools out there. Yoga has been with us for centuries, originating in India in the fifth and sixth centuries BCE. While it might surprise you to see an emphasis on yoga in this book, it really is an ideal grounding tool worth exploring or integrating more regularly into our routines. Yoga is a stress-relieving practice that offers you a container for building strength of body and mind, expanding community, and experiencing gratitude and awe. Because it incorporates breath, movement, chanting, music, stretching, strengthening, balance, and yogic philosophy and meditation, it's most certainly a recipe for a healthy centering practice.

Yoga is versatile and inclusive with its many branches and philosophies, ranging from active to passive practices. In general, active yoga includes strengthening poses, sun salutations, flow, and energizing breath work (e.g., Hatha, Flow, Power Yoga, Anusara, and Iyengar). Passive yoga includes shivasana, calming breath work, child's pose, and receiving adjustments (e.g., yin yoga, yoga nidra). Other practices include Acro or partner yoga (playful and interactive), trauma-informed yoga (feeling safe and attuned to your body's particular comfort and experience), and pre- and post-natal yoga (sometimes even with childcare).

Whatever type of yoga you practice, you are cultivating mind-body-spirit connection, which can help to reset your nervous system amid all of the parent noise and pressures. And it can serve you through life. Yoga classes are very much an intergenerational space and support community building. There are also yoga classes for kids and families. Yoga was transformed during the pandemic when classes continued on video calls with participants joining from the comfort of their own homes. The endearing Cosmic Kids Yoga teacher, Jaime Amor, continues to enchant children with her unique combination of storytelling and yoga.

Whether you're an experienced yogi or brand new to yoga, I'd like to highlight the concept of *drishti*, meaning "focused gaze" in Sanskrit. To help find some semblance of grounding in the midst of today's crises, drishti can be practiced in a way that feels authentic to you.

A Twenty-First Century Anti-Doomer Toolkit for Families

In tree pose, you balance on one foot and bend the other in toward the standing leg. If you feel steady, you might raise your arms straight overhead. The key to this pose is finding a steady point to softly gaze at even as your body trembles. If your mind wanders, then your gaze might as well, and you could fall out of position. However, if you focus too intensely or overthink the pose, then the same thing might happen. Metaphorically speaking, this pose requires you to stand tall like a tree, your body mirroring the strength and steadiness of a trunk, despite what the weather is like around you. You can use focal balance and proprioception, feeling into your body's limbs to tune into focus, strength, and flexibility.

There are other ways to practice drishti as well that don't involve yoga. Staring at the same view outside your office window each day, regularly looking at a meaningful photograph or art, or reading from a particular book of poetry can also qualify. Daily drishti is a powerful practice because you keep coming back to yourself, back to your values, finding a way to sustain that gaze despite all of the Macro and Micro stuff that's clamoring for your attention. Amid chaos, overwhelm, trauma, and intense emotions, drishti can help to counter reactivity, and bring a sense of calm and clarity. The trick is to develop your drishti now so that when things heat up (no pun intended), you have that practice in place.

In Jenny Odell's *How to Do Nothing: Resisting the Attention Economy*, she makes the case that our attention is our most powerful asset and worth protecting at all costs. She recognizes that our attention is commodified and competed for each day by invisible market-driven forces, such as advertisements, web pop-ups, text alerts, and social media feeds. Maintaining steady engagement on important issues is difficult when our lives are regularly punctuated by manipulative marketing, cell phone dings, and screens. In a modern day existential context, drishti might mean how we maintain sight of our mission amid all of the noise or what keeps you from going down the doomer rabbit hole.

FAMILY PAUSE

Pause now to explore drishti in your life. What might be a helpful point of focus for you?

A fun drishti activity you can do with kids is creating a vision board. This can be with photographs, magazine cutouts, images, words, and poems, for example. Or, find a photograph of a special place in the MHW or of a person or place that you look at regularly that you can come back to again and again to ground you.

Dristhi is a place where our attention can settle, and where we can be still in our minds, our bodies, and our hearts. To exercise drishti is both personal—and political—resistance.

Indigenous wisdom: a primary source of grounding. Indigenous communities around the world, who've lived in right relation with the Earth for centuries, are increasingly recognized in the Western world as key players in responding to the climate crisis. Policymakers have come to recognize the importance of Indigenous-led solutions, as they possess traditional ecological knowledge and understand responsible stewardship of natural resources. Educating ourselves, with humility, about Indigenous ways of life is vital, and offers a well of deep wisdom and essential knowledge in how we repair our relationship to the MHW.

Land is viewed as kin, as inseparable from humans, in many Indigenous traditions. The First Nations, Inuit, and Metis have long considered Earth as their "mother." Indigenous Andean cultures honor the goddess Pachamama, commonly translated as Mother Earth, and many believe that problems arise when people take too much from Pachamama. In

Maori culture, Papatūānuku is the land, or Earth Mother, who gives birth to all things. And so on. When the rhythms of our natural world are off, Indigenous communities are often the first to know and are most vulnerable because they depend on these cycles for their survival.

Although there are many important Indigenous concepts that can help guide us and get us back on course in living in right relation with the MHW, I can only share three of them here.

In *Braiding Sweetgrass*, Potawatomi botanist Robin Wall Kimmerer writes about the concept of reciprocity: "Each person, human or no, is bound to every other in a reciprocal relationship. Just as all beings have a duty to me, I have a duty to them."[4] The idea that natural resources are not ours to commandeer, that there is a give and take, a mutual and reciprocal relationship helps to ground and connect us in the web of life that is vaster than the individual.

The Seventh Generation Principle is a Haudenosaunee (Iroquois) concept shared by many other tribes and a reminder that the decisions that we make today will impact our descendants, so we should live as though we are borrowing the Earth from the next seven generations. Again, this requires a reorienting of our focus on self toward a focus on others—on greater selflessness. For an individualistic culture like ours, this will be a major shift. But it's very much in alignment with our care and love for our kids and their futures.

At a conference in Alberta in 1982, native elders, leaders, and professionals of various North American native communities came together to develop an integrative text, *The Sacred Tree*, that encapsulates their shared knowledge and wisdom. It offers a synthesis of native ancestry and traditions that celebrate each person's universal potential, and their interconnectedness with the rest of creation. The Four Directions, as referred to in this book, make up a compass, with East/West/North/South directions, and illuminate a way for us to mindfully orient for the sake of collective resilience and engagement. Here's a short summary of the Four Directions plus a possible application of this wisdom:

- The East represents childhood, innocence, trust, unconditional love, truth, birth, and rebirth. The special gift of the East is the capacity to be fully present in the moment with all senses alert. Perhaps the teaching here is how to remain hopeful and sustain hope while being honest about what's happening in the world.
- The South represents love, sensitivity to other's feelings, physicality, and passionate involvement in the world. The special gift of the South is the discipline of feelings and capacity to express them with compassion. Perhaps this direction's teaching is to reckon with your climate emotions, hold space for others to do so, and communicate mindfully.
- The West represents mystery, power, contemplation, clarity, respect for others, perseverance, and sacrifice. The special gift of the West is self-acceptance. Perhaps this direction can show how self-awareness can prevent burnout.
- The North represents wisdom, complexity of thought, fulfillment, intention, detachment (from strong feelings and thoughts), and balance. The special gift of the North is justice. This direction can guide toward social justice, structural change, and equitable solutions.

The wisdom from Indigenous traditions is vast, and I cannot do it justice in this book. Please visit Appendix F for further resources and learning opportunities and be mindful of appreciating (not appropriating).

The seasons. Around the world, Indigenous communities strive to live in harmony with the rhythms of nature, including seasonal transitions. Their livelihoods have been disproportionately shattered by the effects of climate change, as warming temperatures alter ocean acidity, spur ice melt, and alter food chain pathways. Industry-led mass-scale pollution has disrupted cycles and patterns of nature.

In the Global North, technological inventions have significantly altered our relationship to the natural world in ways that we don't often consider. For example, electricity means that human activities need not cease as it grows dark. Indoor heating and air conditioning create an artificial bubble of comfort for the privileged despite extreme heat and cold. Cellular and internet communications mean that not only do we not need to be next to a person to talk or listen, but we don't need to wait for a letter either. As we've gained efficiency and control over some aspects of our lives, in other ways it's cost us our relationship with the MHW. How can we begin to restore—and heal—that bond of interdependence? And what might we regain by doing so?

To realign with seasonal changes, and restore this connection, you can try small shifts at home. Your kids will not only appreciate it aesthetically, but they may find comfort in the seasonal rhythms and repetitions that play out from year to year, bringing a sense of comfort and safety that comes with predictability. Moreover, your kids will learn that seasonal rituals mark the passage of time, offer space for gratitude and reflection, and will snap us out of MA and experience things through enlivened senses. Your seasonal rituals will ripple out, as they so often do, and can play a role in resetting cultural norms if they catch on. What are some ways you recognize or relate to different seasons? This could be in temperate, tropical, arctic, or other climates. You could also reflect on what it's like when you've visited other regions and experienced seasonal elements like heavy rainfall, snow, or foliage changing colors.

Respect seasonal shifts toward light and darkness. Mammals—and that includes us—hibernate in the winter. Our bodies crave rest, sleep, and comfort food. Lights dim in the evening, we go to bed earlier and sleep longer hours. Quite naturally, productivity decreases and there's a desire to slow down. Here are a few ways to connect and align with the seasons:

- *Buy seasonally and locally.* It makes good sense to eat what grows around you at any given time of year. Try joining a community-supported agricultural (CSA) farm share, where you and the farm both benefit from seasonal harvests, or visit the local farmer's market.
- *Reflect seasonal transitions in your household.* That could mean planting a garden, or changing out winter for summer clothing, or creating a wreath for your door out of what's growing in the garden.

- *Ayurvedic adjustments.* Ayurveda is a traditional medicine practice from India that is more than three thousand years old. Ayurveda divides the year into six seasons and recommends following seasonal routines in diet and lifestyle to maintain health and well-being. You can learn about these and easily implement in your family.

Honoring seasonal transitions is profoundly grounding. It's a way to chart where you are physically relative to the sun and moon, to lightness and darkness, and the natural rhythms of the world. There are many cultures that mark seasonal passages: the Persian tradition of Shab'e Yalda (a community celebration on the longest night of the year), the Japanese tradition of Osoji (a year-end deep cleanse), and the Jewish holiday of Sukkot (harvest festival) are just a few examples.

Through grounding and reestablishing our connection with the MHW, we can create greater stability in our lives regardless of what's happening in the Micro and Macro worlds.

Connected (Hands)

Find a comfortable position, sitting or standing, indoors or outdoors. Take a few rounds of cleansing breaths. Start to notice your breath, your body, and what's around you.

Place the palms of your hands together and rub quickly against each other, creating warmth and friction. After about thirty seconds, cup your hands over your eyes without touching them. Allow your eyes to receive the warmth and then cross your arms on your chest, holding your elbows. If you're in a group, you can create a circle and hold hands or put arms around each other's backs. If you're comfortable closing your eyes, go ahead.

Imagine that you are surrounded by others who care deeply about what is happening in the world right now. You might know them or they might be strangers. As you look around, you feel safe, contained, and supported. You feel the warmth of your hands, a pulsating energy

of anticipation and excitement in the air, a feeling of solidarity and undeniable hope. You trust that everyone is eager and willing to take action . . . notice how you feel. Can you soak in this feeling? Can you carry it with you in your hands?

As you read this section, keep toggling back to awareness in this region, using touch as a way to periodically reconnect.

In an increasingly lonely world—where people are spending less time outdoors, less time socializing, feeling more fearful in public spaces, and loneliness is considered an epidemic—social connection is a fundamental touchstone. A 2011 study of fourteen thousand college students found that today's young people are 40 percent less empathetic than college kids were thirty years ago.[5] One of the researchers suggested this could be because real-life social interactions have declined as computer and screen use has accelerated, leaving fewer opportunities to interact with others and to foster empathy and understanding of other's perspective.

The antidote to disconnect is what Buddhist monk Thich Nhat Hanh called "interbeing," the interconnected state and mutual dependence of all living things. In the section that follows, I explore some ways we can connect with others and how solidarity strengthens our ability to sustain engagement in what matters most to us.

The power of social connection. Did you know that loneliness increases your risk of premature death by at least 32 percent?[6] Social connection might be our very best tool as humans living through a polycrisis—not only in terms of effectiveness and getting shit done, but also for maintaining mental health and longevity. Studies have shown that social connections improve mood, lower blood pressure, and decrease mortality. As human beings, we are primed to connect at an early age as we form deep attachments and bonds to our primary caregivers. Oxytocin is released into our bloodstream during childbirth, prompting an immediate longing to be close, to hug, kiss, bond, and nurse. Our primary caregiver forms a blueprint for our future relationships as we learn early on how to navigate love, trust, and co-regulate.

Let's not forget that humans have evolved as social creatures, living in hunter-gatherer clans and tribes that served their evolutionary fitness. Our very survival depended on our ability to care for others in groups and by not operating as competing family units. As Parent Club members, can we consider ourselves as alloparents and start to reconceptualize our roles in the world today? Not only will this build stronger communities, it will help parents to feel less alone when parenting through today's crises and stressors. In fact, studies show that community ties make the difference in strengthening personal resilience against climate disasters. A 2021 study by *The Lancet* linked community strengths (food security, trusting relationships, diversity) to greater social cohesion and described how these strengths can act as a buffer against the negative health impacts of a climate catastrophe.[7] A 2017 study revealed the centrality of social ties in buffering mental health during climate migration.[8] As more people are displaced as a result of rising sea levels, soaring temperatures, desertification, or ecological destruction, community support systems become invaluable.

Social connection seeds social movements. Have you ever felt the excitement and energy of being part of a group with a shared goal? The term "collective effervescence" describes a feeling of unity and joy that people experience as they come together for a shared purpose. It's uplifting, motivating, energizing, and resourcing.

Recently, my youngest daughter and I experienced collective effervescence when we attended a political event on the beach in San Francisco. A couple hundred participants spelled out a message in the sand with our bodies, which was then photographed by a helicopter above. We waited patiently for the helicopter to get in place, our feet burrowed in the sand, feeling the collective pulse of people coming together to take aligned values-centered action. But to see the photograph a few days later was a whole different feeling: My daughter's eyes widened with astonishment as we zoomed in on the small dots in the photograph that were our bodies standing tall and proud amid a sea of so many others.

To be honest, I'm not sure why we need science to whack us over the head with the obvious: When you have people around you who care, resilience goes up. The comfort that we take from human connection and the need for it doesn't change throughout your lifespan. Heck, plenty of studies show that strong social connections promote longevity, while isolation short-circuits it. How's that for motivation?

A word on hugs: During the pandemic, I published a piece in the *New York Times* about the power of hugging. My husband happens to be a world-class hugger, and I had been thinking about how grateful I was for that amid so much uncertainty and isolation. My kids, too. But this discovery also prompted me to suggest hugging as a way for my clients to connect with friends and family because it's mutually beneficial for both parties. In more than one instance where there was parent-child tension or misunderstanding, I suggested hugging and physical affection as a way to support connection. Where there is tension in a body, or in a relationship, hugging can help. Studies suggest that hugs actually reduce stress, provide an immunity boost, and might even lower blood pressure.

Cultivating strength through ancestral wisdom. Each of us has a family history as well as a cultural legacy trailing behind us. Through time, lineage, and geography, we're part of a dynamic intergenerational web of human life, influenced by generations preceding us, and influencing those following us. Too often we overlook this source of strength because our individual ego dominates. But if you look back at your family history, it's a gold mine of deep wisdom.

Ancestral inquiry can be as simple or as complex as you make it. In addition to trauma and hardship, you can find cross-generational examples of strengths and values that you might learn from or call on at any time. In the list that follows are some ideas to get started in forging cultural resilience.

- Review family photos, documents, written histories
- Speak with an older relative or someone who possesses knowledge of your ancestors and their traditions
- Visit lands, communities, and sites where your relatives spent time
- Create your own family tree
- Write an autobiography or about your family history
- Identify aspects of your family's legacy that you wish to pass on and those that you plan to leave behind
- Learn and practice traditions, rituals, holidays, and ceremonies
- Embark on cultural, anthropological, or historical research
- Share your discoveries with your children (or include them in this process)

Sometimes, in our eagerness to grasp something that seems wise or superior (or due to a lack of cultural exploration), we might overlook our *own* Indigenous ancestral knowledge and traditions. Rather than reach to other cultures for answers, how about starting with your own first? Many of us are unaware of or disconnected from this wisdom. Yet it's vital that we explore and repair our own relationships to nature and ancestry in order to show up wholly and humbly. Take some time to start pondering this, or perhaps set up a weekly date.

Other-care. I came across this term in an article in *The Atlantic* a few years ago. "Other-care" is exactly what it sounds like: taking care of others. However, it's also the idea that caring for others not only helps others but it also helps *you*. Volunteering time at an animal shelter or soup kitchen, mentoring a child, or committing random acts of kindness is meaningful work that ripples out into the community.

In these current times it's important to cement this value into children early on and to practice what you preach. Consistent volunteering is ideal but often challenging. Consider making your commitment seasonal so that you remember that every January or every September you volunteer. In my family, it's a given that we volunteer at a local food bank in wintertime and a beach clean-up-turned-art on Earth Day. Other families volunteer in a community garden in the springtime or at an animal shelter in the summer.

Or keep it simple: pledge to not ignore people struggling in the community around you. Instead of passing by a person holding a sign asking for food, find a granola bar to pass along. Kids are tuned into this in a way that jaded grown-ups are not. There are many ways to contribute, and it's certainly not a shame game. If you aren't doing anything now, let this be a reminder of how uplifting and powerful it can feel for you and your family, and what a difference you can make for those in your community.

Learning about social connection from . . . trees! As I write this chapter, I'm staring at a small grove of redwoods flanking a creek near the Pacific coast. I feel that I'm in good company, welcomed, supported, and safe. I do not feel alone. No matter what is going on in my head or in my body, these sentinel-like presences are here, unwavering. They commonly intertwine their roots to help withstand significant wind and floods, offering physical support to one another, even sharing nutrients. Many trees also emit chemicals to warn other trees of nearby threats by repelling them from the greater area. These redwoods care deeply for each other; it's in their plant DNA. Can you guess where I'm going with this? Trees remind us about what healthy relationships look like. If we can pay attention, then perhaps we can learn how to be a more inclusive, equitable, and compassionate society.

Embedded (Human-Environment Interface)

Take a stroll or hike with your child through a natural setting. Walk slowly and mindfully, attuned to your five senses. Consider talking less in order to drop into the experience. Find a place to sit or lie down in the MHW: a stump, rock, bench, grass, sand, or chair. Or sit in or against a tree. You're welcome to deepen this connection by removing your shoes so that you're barefoot. Feel into the support of the ground beneath you, holding you, that interface between your body and the

ground. Imagine again energetically rooting to the earth through the soles of your feet, or any part of your body touching the earth, breathing gently yet deeply in a fluid exchange. Feel the raw connection that always is and always has been. Earth's vitality is moving through your body, circulating and releasing. Allow your heart to fill up like a balloon as you inhale, and surrender your body to gravity on your exhale. Pay attention to this exchange as you experience your embeddedness with curiosity, wonder, and awe. Close your eyes if you'd like, and spend as much time as you'd like here (you are welcome to set a timer). Flutter your eyes open ever so slowly, calmly taking in what you see, perhaps tracking lightness and darkness as you transition back. Make a motion with your body, a gesture, expression, posture, or movement, and make a sound that expresses how you are feeling. Thank yourself and the MHW for this time.

As you read this section, keep toggling back to your whole-bodied sense of relationship to your environment, as a way to periodically reconnect.

Energetically speaking, Earth's electrons balance out free radical ions in your body. Grounding, or earthing as it's called, can be a powerful way to calm your body as well as tap into a sense of interbeing and awe. Now for a shocking statistic: Did you know that the average American spends approximately 93 percent of their life indoors? This includes time spent both in homes and in vehicles.[9]

As part of my intake process, I typically ask my clients about their indoor/outdoor ratio. Many are surprised by their own imbalance. One depressed client told me that he spends an hour or less a day outside. You can bet that a part of his treatment involved dusting off his old surfboard and catching a wave in the ocean (an old hobby of his). Repairing our spiritual connection with the world around us, and feeling into our smallness amid the vastness of our ecosystems, elicits a sense of wonder that makes us feel good.

A Twenty-First Century Anti-Doomer Toolkit for Families

Connecting with the MHW is a powerful antidote to climate distress. Yes, that's correct: the exact thing that ratchets up your stress and despair is an answer to it as well. Exploring less trafficked and quieter areas abounding with trees, flowers, birds, animals, rivers, and waterfalls, particularly apart from industry and noise pollution, is spiritually replenishing. And sharing this time with your family supports whole family mental health while reinforcing ecological values and reconnecting people to a rich sense of awe and gratitude.

I do want to call to mind an important principle of ecotherapy here. According to a pioneer of the field, Linda Buzzell, there are two levels of ecotherapy. The first level is human-centered nature appreciation and connection that focuses on human benefit. The second level is humans engaging in a more reciprocal relationship and giving back rather than just taking. In other words, maintaining awareness of how you relate to the MHW is essential. Even as we might have good intentions about spending time in nature, are we showing up solely for our own benefit? Or is there a way to offer back to Gaia? (For example, picking up litter or land tending.)

There's great inequity in accessing green spaces and experiences. Restoring fundamental human rights to land and waters is essential, so that you can experience this sense of embeddedness regardless of where you live, who you are, or how much money you have. Meanwhile, the tools that follow are reasonably accessible, although they do require time and ideally access to some form of greenery, outdoor landscape, or body of water. There's a list of low-cost, low-barrier resources and opportunities connecting kids and families to the MHW in Appendix D.

Embedded means experiencing your oneness with nature through your senses. Here are five simple tools to get started:

1. *Just "be" outside!* I mean this quite literally. Level 1 might be this: Whatever you're doing right now, can you take it outside? For example, a phone call, a conversation, a book, a podcast, or a meal. Level 2 might be once you are there, can you carve out time for a more meaningful

connection? This requires shelving the multitasking for a few minutes so that you can really pay attention to your surroundings. You might notice a monarch butterfly fluttering through the air, feel the sensation of a cool breeze against your cheek, smell freshly cut grass, or hear the sound of trickling water. Spend time breathing, moving your limbs, and smelling things, and listening, looking, touching, and tasting.

Simply giving yourself permission to be fully present in the natural world enlivens you with a vitality of spirit that is difficult to experience when sequestered within four walls. Keep in mind that kids intuitively love exploring the natural world: examining interesting bugs, catching tadpoles in a stream, playing mercilessly with sticks, or climbing trees, for instance. Taking a pause to be in nature will reverberate throughout your family, money back guarantee. For Level 3, try a nature-based practice or ritual. This just means something that you do with regularity in MHW. Of course, you can do these alone or with kids. Check out my website for some ideas.

2. *Doctor's orders!* Many in the medical community are starting to embrace the healing potential of the natural world. Some US psychiatrists are writing "park prescriptions," as in a prescription that is not pharmaceutical but rather is nature-based. To help improve symptoms of anxiety, depression, ADHD, and PTSD, patients are now prescribed to spend more time among trees, birds, insects, and the elements.

In Japan, forest bathing (*shinrin-yoku*) is a common intervention prescribed to help lower blood pressure, reduce stress, boost mood and concentration, and improve sleep. Patients may spend a couple of hours walking or sitting in a forest with medical staff and equipment present.

But to really understand the forest from the trees, let's dig deeper. Trees emit chemicals called phytoncides, which have antibacterial and antifungal properties. When people inhale phytoncides, their white blood cells count increases, boosting immunity, reducing inflammation, and improving cardiovascular health and brain self-regulation. A study by Stanford University indicates that spending just ninety minutes of calm, reflective time in nature decreased neural activity in the brain's prefrontal cortex, which is responsible for rumination and negative thought patterns.[10]

Also good for your mental health is regular skin contact with soil, such as gardening or walking barefoot. Studies suggest that microbes in the soil might stimulate serotonin production, resulting in feeling happier and calmer. Many gardeners will attest to this shift anecdotally. To read more about this, check out *The Well-Gardened Mind: Rediscovering Nature in the Modern World*, in which psychotherapist Sue Stuart-Smith writes about gardening and mental health.

3. *Cultivating creativity and healing—all in the great outdoors!* From *en plein aire* painting and journaling to earth-based and Indigenous arts (e.g., basket weaving, fiber arts), there's no shortage of combinations to take creativity to an exciting next level in this realm.

Eco-art therapy is an exciting field that combines ecotherapy and art therapy. I cofacilitated a group a couple of years ago that blended forest bathing with spontaneous art-making. I heard comments like: "I forgot how easy it is to just make art . . . and that I can share this with my child." And: "I forgot how important it is for me to just spend time outside." Participants found nature allies and incorporated these into artwork, including leaves, twigs, and even an eggshell that had fallen from a nest.

On an overnight yoga retreat—my first time away after my eldest daughter was born—I had the honor of meeting Joseph Cornell, a well-respected nature educator and author of *Sharing Nature with Children*. He led a dynamic workshop where we imagined embodying a tree's perspective as it changed through the four seasons of a temperate climate and physically assumed a tree's perspective in a forest as ways to relate to the MHW.

4. *Bring nature indoors.* One of my clients told me that he has over thirty plants inside of his city apartment. Throughout our home, I've placed river rocks that my husband and I have independently collected from our various journeys throughout the world. There's a small pile of them in the corner of our shower, and I love watching them get wet, and rearranging them with my foot into new formations. Same with a shell

collection housed inside of a large abalone shell on our shower shelf. A colleague of mine practices ikebana, a Japanese flower-arranging art, where you select specific blooms and greenery to evoke feelings. These are all easy ways to bring nature indoors.

Note: Since becoming a parent and eco-aware therapist, I've begun giving greater consideration to taking nature allies out of their home environments. Taking or collecting nature allies can profoundly alter and denigrate an ecosystem, and it's important to be thoughtful and intentional about what is taken out or disturbed in an ecosystem. What is gained? What is lost? How much is being left behind? We must teach our children as well.

Creating a centerpiece or art piece that marks seasonal transitions embeds nature into your communal living space. In the fall, depending on where you are, maybe it's pumpkins, squash, or sunflowers. In the spring, maybe daffodils, marigolds, or cherry blossoms. Kids love to mark the changing of the seasons, and they can help with picking and arranging flowers and other items. Bonus: You can draw or paint these arrangements!

5. *Awe.* Sometimes, when I feel confined by a political moment or I'm spinning out mentally, I step outside on a dark night to view the night sky. Gazing at twinkling stars or the phase of the moon, or at a planet for long enough to imagine its size or what is happening in its depths, reminds me of my smallness. I think: *Oh yeah! I'm but a speck on this planet and my lifespan is a blip!* I do this with my kids while we are camping, searching the sky for shooting stars or the Big Dipper. Finding awe in the living world conjures humility and it keeps us honest.

As a new mom, I remember reading somewhere that you can't force a baby to sleep—you need to let the quiet of the room and baby's own sleepiness gradually overtake them. That is an apt description for what happens in nature too. You don't force anything to happen; rather, just by being present, you gradually feel yourself absorbed into your surroundings, until you experience a shift in consciousness and perspective, maybe even awe. Rabbi Abraham Joshua Heschel said, "Our goal should be to live life in radical amazement . . . get up in the morning and look around at the world in a way that takes nothing for granted. Everything is phenomenal; everything is incredible; never treat life casually. To be spiritual is to be amazed."[11]

As this long toolkit chapter wraps up, my hope is that you and your kids feel liberated to try something new or reconnect with practices that have served you in the past. Just as the body functions as a whole, so do these tools. They do not stand in isolation. You and your family can find yourself using a couple of tools at once and reap the positive benefits (e.g., eco-art therapy, yoga in community). More than that, practicing these tools regularly will cement them into your family culture and create positive habits. Be sure to keep up these centering practices as you finish the rest of this book, and beyond!

To help you integrate all of these varied tools in the EBM, I recommend creating your very own self-care acronym. These can be kind of silly, both the process of coming up with an acronym and sharing it with others, so prepare to have some fun. The self-care acronym is most definitely a kid-friendly tool—just keep it short and simple.

SELF-CARE ACRONYM

Think of the things that you love to do or aspire to do that bring calm, joy, and balance into your life. Write down your list on paper, then put the letters in an order that you'll remember or that makes sense to you, or form something memorable. (For example, FOMO stands for "fear of missing out.") Write it

out on a sticky note, or get fancy. Be sure to post your list somewhere where you can view it—like near your desk, on a bulletin board, or on the fridge—and check it regularly.

Here's mine, which is funny because of its random complexity: YMCA NIB QFR.

Yoga
Music
Creativity
Acupuncture

Nature
Incense
Bath

Quiet
Family/friends
Retreats

When times get tough, the self-care acronym is a vital lifeline. For kids, keeping it simple is the key. Maybe their list is just "DSB" (dog/sleep/basketball). Earlier this year, I visited my aunt, who was caregiver for my dying uncle. As the sole therapist in my family, I knew my job even before I touched ground at her home in Boise, Idaho: to help her regain some balance and calm as she was exhausted, overwhelmed, and burned out from all the health-related decisions she was juggling amid the grief of losing her partner. I took her through the self-care acronym exercise. She realized that she'd forgotten some of her basic needs and wellness activities: mindful eating, petting cats, singing, and walking in the fresh air. She posted her self-care acronym on the fridge and it helped her to remember to do these things and to recenter.

A Twenty-First Century Anti-Doomer Toolkit for Families

To close out this chapter, you'll be doing an awesome exercise that explores your relationship with the MHW, as well as with your ancestors. This is an adaptation of Dr. Thomas Doherty's Eco-Identity Timeline[12] and a helpful integrative exercise.

CREATIVE EXERCISE

AN ECO-ANCESTRAL IDENTITY TIMELINE

Materials:

Option 1
- Paper, poster board, recycled cardboard or side of a box
- Markers, gel pens, colored pencils, Sharpies
- Any fun supplies to decorate with (e.g., collage, stencils, stamps, stickers, glitter)

or

Option 2
- Find a long (dead) tree branch, natural stalk, or wreath
- Cardstock, any type of paper, origami paper
- Markers, pens
- Hole punch
- Yarn or string

Setup:

On scrap paper, brainstorm a list of meaningful moments in your life in nature, whether positive or negative. Parents may need more time to do this than kids. To help guide young kids, you can just ask them to name or write down some of their favorite memories in nature or with animals. For teens, ask them to think about some of the harder or less enjoyable memories as well.

If you haven't already, think or research about your ancestors' relationship to the MHW, and any practices, traditions, or events that might be important to include. This is an opportunity to teach your kids about your family culture, resilience, and even loss (if it feels developmentally appropriate).

Feel free to create a sketch, rough draft, or concept design first as you think through your ideas.

If you're doing option 2, then you'll want to cut out many small same-size papers and punch a hole into them, as you will be tying these onto the branch, stalk, or wreath.

Prompt:

Create a timeline that shows your relationship to the MHW and spans your lifetime, if possible including your ancestors' relationship to the MHW. If you are doing this 2D, then think if this will be linear, a spiral, or some other shape or metaphor to chart this. If you are doing this 3D, then you can write onto each paper a particular memory, and when you are done tie these onto the branch or wreath in an order that makes sense to you.

Process:

As you create, notice without judgment any feelings, thoughts, or sensations arising in your body.

Product:

When you're done, step back and reflect on your creation. Spend some time with the reflection questions that follow, and journal or share your creation with your family or others.

A Twenty-First Century Anti-Doomer Toolkit for Families

Reflection questions:

- At what point in your life did you feel most connected to the MHW? The least connected?
- How has your relationship changed over time with the MHW?
- How has your family of origin shaped your relationship with the MHW?
- How might you cultivate this relationship with your own kids?
- Did any emotions or sensations come up?

Note: If you feel triggered by this exercise, please visit the trauma resources in Appendix C.

SEVEN

How to Talk to Your Kids Amid a Brewing Existential Sh*tstorm

"At the end of the day, I want to protect my kids from all of this sh*t."
—Parent Club member

The media has sounded alarms warning us that talking about climate change may stoke anxiety and existential dread in our children. But such absolutes do us no favors. The question should not be *Do we talk to our kids at all?* but *How can we talk to our kids about this?* After all, our kids are going to inherit our world, and we want them to be prepared for that eventuality. The art of talking to our kids about tough real-world topics lies in the nexus of three things: *what* we choose to share (or not share), *how* we share it (tone and demeanor), and *when* (timing).

Picture yourself at the end of the day, kicking back and scrolling through the news, when you spot an article about catastrophic flooding in a nearby region. In a nanosecond you skim images of displaced families and see the death toll and fear grips your heart. You think: *OMG. This could happen to us*. Feeling anxious and frightened, you start ransacking the house, prepping a "go bag," and talking about it indiscriminately in front of your kids, your emotions spilling out every which way, in a deluge-like fashion.

Guess what? Your kids are going to feel your stress big time. They'll wonder what's up with mom or dad and begin worrying about you. They'll absorb your raw, unmetabolized emotions, grow uncertain about their own safety, and feel more unsettled by the pandemonium at home.

Now, imagine that same scenario, but instead of moving into high gear after reading the news article, you take some minutes to yourself, reflect, journal, or talk to someone, allowing yourself to work through your emotions before involving your kids or taking action. When you start prepping (perhaps later that day or some days later), you inform your kids as calmly and matter-of-fact as you can, "In case you're curious, I want to tell you about what I'm doing. There's been some storm flooding, and it made me realize that we should be prepared in case that ever happens here. So, I'm going to pack some supplies and make a safety plan that I'll go over with you just in case we ever need it one day. If you have any questions or want to talk about it, just let me know."

The message conveyed here was calm, organized, and confident. Your kids will interpret it as mom or dad is making sure we are safe, just in case. While that news about flooding might get them thinking more seriously about such a threat, they'll still internalize a message that they're safe right now, that their parents can handle this, and that they can to. And, if they feel unsettled, they know they can talk to you about it and that the topic is not off-limits.

Talking climate or other tough real-world topics with kids confounds many parents, who struggle to balance appropriately sharing and preparing their kids with the desire to protect them from discomfort and suffering. While excess details and unfiltered emotional responses can freak kids out, not broaching the subject of climate change at all does not help them build resilience or adequate awareness, both of which will be critical to their survival. But just as you've prepared your kids for future scenarios like what to do if a stranger approaches them, if someone offers them drugs or alcohol at a party, or if someone makes unwanted sexual advances toward them, this is also a must-have conversation that will better prepare young people for the future.

What's the alternative, anyway? Silence and secrecy does no one any good; it only creates a fragmented family culture and induces shame, loneliness, and/or fear. According to Buddhist eco-philosopher Joanna Macy, when it comes to kids, silence conveys fatalism and indifference, and it reinforces repression. What we choose *not* to talk about, what we consciously or unconsciously omit, speaks volumes. So it's best to make this a choice rather than something that just sort of . . . happens (or not).

In the psychotherapy world, it used to be commonplace for therapists to use a "colorblind" approach that treated clients "the same" (whatever that means) regardless of their race or culture. Imagine the experience from a client's perspective, and how that failure to acknowledge or ask about culture can create discomfort and tension. In today's cross-cultural counseling, professionals bring curiosity early on about a client's culture, religion, gender identity, sexual orientation, disabilities, or neurodivergence. Doing so increases trust in the therapeutic alliance and a sense of being seen and understood—and allows for the open navigation of those topics in therapy.

The point that I'm making here is that omissions in communication ratchet up intra- and interpersonal tensions. No matter how afraid we might feel, just making an attempt at a difficult conversation with a child might offer them some relief, something tangible to grip onto amid all of the uncertainty. As the saying goes, "A problem shared is a problem halved." (Incidentally, the second part of that saying is: "A joy shared is a joy doubled.")

As you read on, you'll discover some tips and tricks for having these difficult conversations with your kids so that you can go into them feeling more confident and empowered. I'll start off by covering key developmental considerations, highlight several strategies, and close by touching on the importance of preparing kids for intergenerational conversations.

How Do I Know If My Child Is Ready for This Topic?

Whether your children are toddlers or teens, you must be sensitive to their developmental stages, needs, challenges, and timing. And while it might be tempting to organize these stages around age alone, it's important to know thy child—which is to say, to know what type of and quantity of information they can reasonably tolerate and find the right approach based on that. For example, there are many tweens who are acting like teenagers (the hair, the fashion, the eye rolls) and might even present as teenagers cognitively, but they are not emotionally ready for teenage-level communication. Or, for example, you might be the parent of an eight-year-old child struggling with a learning disorder or social skills who might be better met with very simple messaging corresponding with a younger age. Some kids do better when they have information up front. Others, such as highly sensitive children, might need a more gentle and incremental approach.

Kids will unconsciously employ their own defensive mechanisms at times to help manage their emotions. For example, when my oldest was eight years old, there was a fatal shooting of a recent high school graduate that occurred in our neighborhood. A neighbor who is an ER doctor and lived close by attended to the victim when she heard gunshots and cries for help outside her home. My daughter heard the story from her friend, and the same day heard a ghost story from another friend. She mentioned both stories to us, and then said that she preferred the ghost story. On some level, she wasn't ready to process what that other story conveyed: that people do bad things, and violence can happen close to home. Children will find ways to cope with difficult stories and experiences, and it's our job not to assume or project onto them but rather to stay with them because they might be making sense of things just fine.

The table that follows offers some examples of developmentally sound dialogue and is intended to give you a sense of what a kid may need at different developmental stages. We'll build off this framework as the chapter progresses.

How to Talk to Your Kids Amid a Brewing Existential Sh*tstorm

How to Talk to Kids About Climate Change Cheat Sheet

Age	School Level	Developmental Attributes	Climate Strategies	Sample Climate Talk
2–4	preschool	sensorimotor exploration, play, exploring autonomy and purpose	• building sense of empathy and care for the natural world • using five senses to explore the natural world • experiencing awe and interconnectedness	"Ooh, look at that beautiful orange and black butterfly!" "Do you smell that? That's the smell of leaves and soil mixing together after the rain." "We are all a part of a web of life: plants, animals, people . . . all that is around us."
5–8	early elementary	sense of mastery and competence, eagerness for basic concepts and facts, sense of belonging	• reading and listening to stories • learning about fairness and justice • cultivating curiosity • experiential learning • learning about ecological systems and cycles • interest in groups, teams, community projects • "no fear before fourth grade"	"Let's study the life cycle of a butterfly!" "Do you know what happens underneath the forest floor? Trees and groups of mushrooms—called mycelium—communicate and work together to break down animal and plant matter to make healthy soil, and even protect each other from harmful insects." "The Earth is getting warmer each year because of pollution from factories, cars, and airplanes. We have the power to stop this and let the Earth heal. We already have clean energy like wind and solar power that doesn't hurt the environment. Who's seen a wind turbine or a solar panel?"

(continued)

How to Talk to Kids About Climate Change Cheat Sheet (*continued*)

Age	School Level	Developmental Attributes	Climate Strategies	Sample Climate Talk
9–12	late elementary to middle school	socio-emotional skills, peer acceptance, academic achievement, being part of a tribe	• science fairs and research-based projects • action-oriented, nature-based learning (e.g., field trips, beach cleanups, hiking, community gardens) • understanding values • sense of personal and collective responsibility	"Let's plant a grove of trees and shrubs that attract butterflies. We can create our own green corridors." "Beneath the soil, there's a whole network of fungi, or mushrooms, that help decompose plant and animal matter and make better soil for forest health. Let's take some soil samples and compare them under a microscope." "Climate change is happening because some activities that people are doing are making Earth hotter. It's not fair that the people who contribute the least to climate change are the most affected. Many folks are working hard to make sure that those most affected by climate change are getting the help and resources that they need. Each one of us can make a positive impact in the world."

How to Talk to Your Kids Amid a Brewing Existential Sh*tstorm

How to Talk to Kids About Climate Change Cheat Sheet				
Age	School Level	Developmental Attributes	Climate Strategies	Sample Climate Talk
13 & up	high school and above	identity, self-acceptance, personal growth, abstract thinking, critical thinking, problem-solving, experimenting with risk-taking, freedom within limits, new responsibilities, social milieu	• cultivating a personal relationship with natural world • environmental stewardship • developing opinions and exercising voice • understanding environmental justice, systems of oppression, intersectionality • exploring climate emotions • advocacy and activism	"I know that you're worried about the decline of monarch butterflies. What about researching some things you can do to help protect them—possibly in your backyard, as a part of a group, or with volunteer organizations." "Do you know about soil health and regenerative practices? I saw a film about them, and it's pretty amazing how soil can drawdown carbon from the atmosphere. Want to watch it with me?" "I know that you're very worried about climate change and what this means for your future. I get it, and I'm right here with you. It's not fair that the people suffering the most are the ones who've contributed the least to the problem. Do you want to attend the march this weekend? We can go together—unless you have other plans."

The What, the When, the How, and the Framing

I kicked off the chapter with an example of having a difficult conversation with your child. Over the next few pages, I'll be covering these fundamentals in more detail. As you read, please reflect on your own child's unique considerations and what might be the best way to approach them.

1. **The What: Serve the Goldilocks porridge that's "just right."** As referenced earlier in the book, like Goldilocks, you must find the porridge that is just right for you and your child. For younger kids, your job is to monitor the inflow of real-world intensity to the best of your ability, much like setting an IV drip to the right setting. Information overload may lead to inundation/overwhelm/anxiety, while keeping silent might eventually do the same.

Start with a warm-up conversation; you can always build on it as you sense into their comfort level/response. Watch them for cues. Are they curious and open? Or do they shut down, mouth off, or walk away? Just as you'd never force-feed porridge to an infant, you wouldn't verbally force-feed a child or teen. Think about feeding your kids bite-size portions of reality, just like you once fed them bite-size portions of food, pausing and giving them time to swallow and digest. Don't pile it on, or they might choke! If you find yourself walking into a power struggle, back away slowly and regroup. No matter how the conversation goes, just remember that you've planted a seed that you can return to and build on over time.

It's important to consider where a young person is at in terms of their conceptual understanding and exposure to this topic. A safe place to start is by asking them what they already know about a difficult topic. After reflecting back that they know some things, remind them that you are there to talk about or troubleshoot their thoughts and feelings with them.

Half of the battle is trusting that you know your kid's developmental stage, disposition, and personality well enough to navigate the window of not-overwhelming-yet-not-underpreparing. It is actually a pretty forgiving space and is wider than you might realize.

How to Talk to Your Kids Amid a Brewing Existential Sh*tstorm

A rule of thumb when talking to kids, as I've heard it described by a colleague, is to "set it up or clean it up." You have two choices: help kids to understand issues by framing or presenting them ("setting it up"), or wait for kids to hear about an issue somewhere and bring it to you. Then you can help them to better understand the topic, reverse erroneous thinking, or possibly reframe ("cleaning it up"). In other words, you can adopt a wait-and-see approach or be more proactive about it from the get-go. Keep in mind that sometimes damage control is more time-consuming and confusing for kids.

Another balanced Goldilocks strategy is called The Sandwich. This simple approach is to carefully curate what you share in an A-B-A format. A equals something positive, comforting, a given. B equals something tough, perhaps some constructive feedback, or a tough topic or news item. The second A equals ending with some closure, positive support, or an idea or plan. The Sandwich, as we know, can be made up from many different ingredients and recipes, and is very versatile. But having those two slices of bread helps kids digest a tougher middle.

As you venture forth into these conversations, rest assured that if you're intentional and real with your kids as well as conscious of what's developmentally appropriate, then you're as prepared as you ever need to be. At the end of the day, by initiating these tough conversations, they'll grow up trusting that some grown-ups are paying attention, listening, and responding. And that will make a difference in shaping their worldview.

2. The When: Timing is everything! If your child had a hard day at school, is feeling stressed about upcoming exams, or is already moderately anxious or down, then consider the *when*. Recall the example at the start of the chapter, discussing flood evacuation plans when your child is already considerably distressed might not be a good idea. If not in immediate danger, you can always talk about it in the future when they're in a more stable place.

Sometimes you might be prepared for the talk, but your kids clearly are not. So pivot, my parent friend, pivot! Kids' emotions are like the weather: hot, cold, stormy, bright, cloudy, clear, gloomy, windy, dark, blue skies, hazy, and everything in between, and they can change in a matter of moments. You might receive irritated snarls, defiant gestures, yawns of boredom, shutdown body language, anxious or restless energy, escalation, and who knows what else. Notice these responses and do your best to interpret them as you decide whether to forge forward with your conversation. Here's a brief summary of verbal and nonverbal cues to help equip you in advance in identifying whether it's a good time to approach (or pivot, if needed):

Signs That Your Child Is Open and Ready		
Mood Signals	**Body Language**	**Tone**
calm	physically affectionate	bright
interested	upright posture	animated
happy	faces you	chatty
upbeat	makes good eye contact	relaxed
bored	nods, smiles, or laughs	normal volume
confident		normal rate of speech
curious		
positive		
easygoing		

Of course, some of these emotions or body language can be directly related to the emotions they are experiencing as you discuss the topic. It's important to try to determine if their reactions are their way of processing emotions or if it's a timing issue. You can always be direct by checking in to see if what you're sharing feels too intense. You might say, for example, "Does what I'm sharing with you feel okay, or is it a little too much?" Either way, you want to make yourself available to talk and provide support. If they're not ready to talk, that's perfectly fine. You can let them know that you can talk about it another time.

How to Talk to Your Kids Amid a Brewing Existential Sh*tstorm

Some Signs That Now Is *Not* the Time		
Mood	**Body Language**	**Tone**
anxious	wide eyes	whiny
depressed	sleepy	irritated
angry	avoids physical contact	volume louder or softer than usual
distracted	averts eye contact	
lonely	restless	talking more rapidly than usual
ashamed	pacing	tense, on edge
hopeless	hyperventilating	lethargic
insecure	slumped posture	
irritated	furrowed brow	
cynical		

Be gentle with yourself, and patient, because this is a learning process. You're not expected to intuit the exact approach needed with your child at every age and stage. Just do your best. Follow your kid's cues, and use your intuition and parent savvy to adjust the formula as needed. Perfection, remember, is an illusion and a trap. If it goes south, take a beat and regroup, but do not lose hope and abandon this course completely. You can usually find an understanding ear in the Parent Club community to help keep your spirits up.

3. The How: The "three V's" (Voice, Validate, Vision). Each week, I bear witness to the evolving contours of youth culture. I hear stories of intense academic pressure, relentless social drama, and flagging self-esteem; worry about US/world politics, war, gun violence, and climate; and overheard racist comments and family tensions. In short, the sheer overwhelm in the face of modern life and its multitude of stressors and expectations. I often joke with parents by telling them the main thing that happens in therapy is that a young person has a safe place to gab about whatever's on their mind with a trusted adult who is far removed from their own life. Seriously, it's all about the relationship!

Over time, I've naturally adapted my own Goldilocks recipe that allows youth to feel safe enough to trust me and empowered enough to speak up about what's on their mind while keeping the family system. When things go well, they may feel safe enough to broach more tender issues that they avoid with their parents or feel ashamed to bring up around others. Of course, sometimes I'm asking pointed questions at just the right time (that's what I'm paid for). I'm then in a position to support them in (a) feeling seen or heard, (b) developing coping strategies, (c) communicating some of these issues with their parents or others, and (d) resourcing them in other ways. Sometimes our work together bolsters a child's confidence in addressing a problem on their own, and we might prepare for a difficult conversation in advance. Other times we'll have a family session, which is emotionally demanding but that is often fruitful labor. I consider myself a kind of intermediary of sorts who can support vital communication but with the eventual goal of fading into the background so that my clients can get on with their lives and learn to do it for themselves.

I'm going to reveal my very simple approach in a moment, but first, a word on meeting kids where they are at. I cannot emphasize enough the importance of this. Relationships are not formulaic; they consist of an incredibly complex set of verbal and nonverbal exchanges, including tone, expression, and body language. When it comes to what makes a child feel comfortable and supported, it varies. Sometimes, it can take me weeks or even months to get this right! Other times, I need to reset our connection after a vacation or break. But if I watch and listen carefully, then I'll know what to do. The same applies to you with your child.

I started meeting with a teenage client who spoke in a whisper, so much so that I had to lean forward in my seat and turn down the noisy air filter to hear her. Over time, by gauging her responses to me, I learned to turn down my own volume and animated gesturing, to slow down and soften my delivery, to listen more, and to allow for longer pauses. Eventually, she began to trust me and share bits of her struggle. Because I met her where she was at, she began to feel safe and accepted. This allowed us to move much deeper and work together.

PARENT PAUSE

Picture your child for a moment. What is their sweet spot, their place of comfort? What types of word, actions, and activities support them? Enjoy conjuring up this image of them.

Now consider: What causes them to move away from that place of comfort into greater vulnerability? What are their edges, and how can you recognize them?

The first V: Voice. By "voice," I mean creating the optimal conditions for a child to speak their truth about their experiences, including sharing thoughts, beliefs, opinions, and emotions. Providing kids with a year-round welcome mat—whether they take you up on it or not—communicates trust and support. You can't assume or expect your kids will just come to you if they need to. I know I sure didn't go to my parents as a child, especially in middle and high school. How about you?

Here are some ideas about how you can cultivate voice with your child:

- *Adopt an open-door policy.* In your own words, make sure you're communicating something like this: "You can always come to me to talk, no matter how hard it feels or afraid you are. Even if I don't agree with you or see things the same way, I will always appreciate your honesty." Be sure to reinforce that when they do come to you; for example, "I'm really glad you brought this up."

- *If talking is hard, try using another tool.* You could try a 1–10 check-in scale to see how they are feeling (1 = awful, 5 = neutral, and 10 = amazing). Another tool is using a printout of emojis so kids can point and identify their feelings, or getting an internal weather report (i.e., sharing emotions using weather language such as "clear skies" or "grey and drizzly"). You could also invite kids to play music that captures their mood.
- *If kids are open to creative expression.* You can invite your child to write a song, a poem, or a fictional story. Have them act out how they are feeling charade-style and you can be the one guessing. Sometimes prompts work well, such as "What do you feel most worried about right now?" You could put a few prompts on piece of paper inside of a jar and have everyone pick one (or use Table Topics or a similar question game).
- *Hold space for a monthly "state of the union" check-in for the family.* This is more of an umbrella family meeting, but there's always an opportunity to introduce special topics. Many kids are familiar with the "Rose, Bud, Thorn" check-in, in which you name something you feel happy about (rose), sad about (thorn), and something coming up soon (bud). The point is to regularly offer a space to check in on how family members are doing, to model healthy sharing, and normalize it.

That teenage client I mentioned earlier who initially spoke in a whisper is graduating from high school and set to enter college in the fall. Her growth over the last couple of years has been remarkable. Not only does she speak in a louder, more confident voice, but she also exercises her voice now in leadership positions both in school and the community. For her, it was the first V, voice, that made all of the difference. I got out of her way and gave her time to find and practice exercising her authentic voice. Meanwhile, she had opportunities to strengthen her voice in new situations over the summer. Using her voice built up her confidence and she stopped worrying so much about what others thought or trying to please them as much.

Here is sample dialogue of how voice operates:

Parent: I know that we don't often talk about how it's so hot where we live, or that the hurricane warnings can be scary, but I want you to know that you can always talk to me about your feelings about this stuff.

Child: I'm fine, Mom.

Parent: I hear you, but I'm wondering if there's anyone you talk to about this with?

Child: [*Sighs*] My friends.

Parent: Oh, that's good to hear. Is it helpful to talk to them?

Child: Sometimes.

Parent: Well, just know that I'm around if you ever feel like talking about it. This might surprise you, but it can be a relief to talk about your emotions sometimes. They can feel smaller once they're out of you.

Child: Okay, whatever, Mom. You worry too much.

Parent: Well, I love you a lot, what can I say? I just want you to know the door is always open.

Notice that it might not seem like much progress was made in this conversation, but there's still a takeaway for your child that you care and that you are there. That's enough.

The second "V": Validate. As important as cultivating voice is, the sheer act of validating your child's emotions is just as important. How many times have you heard "Stop trying to fix me!" or "I don't need your help—I'm fine"? Those statements might both be indications that validation is in order to repair your relationship.

"Validate" means to actively listen while refraining from reactive comments, judgmental remarks, and facial expressions. But the second important piece is reflecting back what you hear to make sure you are understanding correctly. Stay with your child's experience rather than share your own or project your stuff onto them.

Just the other night, we sat down as a family to watch a nature documentary. My nine-year-old was upset by footage of two humongous yaks attacking each other as part of their turf wars, their horns gnashing against each other as they charged each other over and over again on a mountainside. She let us know that she was upset by shrieking out "Oh no!" and averting her eyes under a blanket. My husband's immediate, unfiltered response was: "Oh, come on . . . this is nature, this is how things work."

Later on, we unpacked his reaction to her as it was far from validating her feelings in the moment. Some of what was behind his reaction was his expectations for her as the oldest child, as well as his own scientist-skewed point of view. In hindsight, he recognized that our daughter is entitled to her emotions and particular sensitivities, and that he would convey his perspective in a less reactive, less critical tone the next time.

Validating is not always easy to do. Indeed, sometimes it can feel directly at odds with our own parent agendas as we seek to empower and prepare our kids for life. But when we attempt to mold kids rather than stay with their experiences, we miss them. Here are a few pointers to get you started.

Attunement, attunement, attunement. It's easy to get triggered by the content of what your child shares. We want to protect their innocence and we love them so dearly. We want to jump in there and start getting to the bottom of it. But your reactivity will beget more reactivity and escalation on their end or lead to shutdown and withdrawal. This is not what you're going for. If you can stay attuned to their emotional response, your child will feel seen, held, and understood. Nonverbal cues are powerful. These include making good eye contact, nodding, murmuring understanding, and conveying warmth in your expression and body language. These simple exchanges are foundational. As a parent, you act as a co-container to help your child hold their emotional stuff. You adapt and attune, attune and adapt.

How to Talk to Your Kids Amid a Brewing Existential Sh*tstorm

Your child is not a fixer-upper. Our culture is big on efficiency and short on time, and I think that's why people are always trying to "fix" other people. It's fast! It's effective! Listen to me, trust me, I know! More often than not, your kid is not asking you to fix them. They just want to be listened to, respected, and supported. They want to know that their thoughts and feelings matter. If your child says, "I'm feeling really depressed that the coral reefs are dying," they don't need to hear a response like, "You don't need to worry about that at your age."

A more skillful response would be: "Oh, honey, I'm with you. It's really sad. I'm glad you told me that you've been thinking about this. Let's learn more about this together and see what we can do to help." Children, like adults, need to feel validated in their very real and understandable grief. The devastating truth is that grown-ups are handing kids a devastated environment. Vitamin V—validation, that is—is a healthy response that promotes connection and helps kids to better articulate and understand their emotions. Hold off on your decrees, ideas, and solutions until it's the right time.

Radiate empathy. If you look at your child with annoyance (guilty), check your phone with impatience (guilty), sigh (guilty), or appear uncomfortable or distracted (guilty), then you'll miss out on real opportunities to deepen connection. Before sitting down for a tough conversation, it's always wise to take a moment for a parent pause—even just three intentional deep breaths. Try it now: 1 . . . 2 . . . 3 . . .

Children are constantly searching for clues in your face or behaviors as to what you're thinking (e.g., *Uh oh, Dad's turning red again, he's going to yell . . . Uh oh, mom's driving fast again, she's angry*). By reassuring them with a warm, empathic expression, a lowering of your voice, an arm around their shoulder, or a hug, you set the tone for a positive exchange. Empathy radiates and often begets more empathy in a positive feedback loop. Of course, dismantling your own defenses exposes your vulnerability and you might feel triggered yourself, but you'll have plenty of tools for that by the time you finish this book. In these situations, one foot goes into the emotional muck with your kids, while the other remains firmly planted outside of the muck on solid ground so that you don't get swallowed up into a mudhole. This strategy allows you to join with kids by validating them, yet remain anchored in your core values, beliefs, and parasympathetic nervous system.

Here's an example of validation in action.

CHILD: The world's going to be over soon because of greedy people who only care about themselves.

PARENT: [*Inhales, exhales*] You're feeling pretty bleak about the future, huh?

CHILD: Yup . . . [*parent waits rather than responds here*] . . . the people in charge keep making excuses. Nothing ever gets done. What does it matter if I do my homework or not if the world's gonna end?

PARENT: I really get the frustration, I do. You're worried about the state of the world when you shouldn't have to take that on at all. I agree—it's not fair. I want you to know that I care, and there are plenty of other adults who do as well, and we're working hard on this. But I want to keep hearing your thoughts and talking about it, okay?

CHILD: Yeah, I know, Mom . . . but talking isn't going to solve this.

PARENT: What do you think will help?

CHILD: If the government actually does something about it.

PARENT: Would you like to talk more about that? I have some ideas about how we can help with that.

CHILD: Yeah, I guess, but I'm already involved in my school club.

PARENT: Well, great! If you want me to help out at your club, I'm always game for that.

Once your child feels validated and secure that you get them at least to a degree, then you can move on.

The last and more tangible "V": Vision. "Vision" is the actionable and roll-up-your-sleeves "V." It helps kids find their path, chart a course on their terms, build community, and enhance resilience for the long haul.

How to Talk to Your Kids Amid a Brewing Existential Sh*tstorm

Here are a few nuggets for supporting your child's vision.

Help them to weigh their options. Sometimes kids need help in making a decision or determining a best course of action. There are many ways to explore life choices. One way I do this is by examining options through two lenses: One, what does my logical brain say? And two, what does my gut say? (Sometimes, connecting to the gut takes a bit of prework, like going for a walk, meditating, or journaling.) Both are helpful places to explore from. Another approach I've taken is painting an imaginary picture of how the two (or more) scenarios might play out, kind of like versions of the final showcases in the TV game show *The Price is Right*. Which feels like the better option for you right now: What's behind door number 1 or behind door number 2?

You can help your child see the pros and cons of different paths and preparing to navigate potential barriers. If a child feels like they need to take some sort of action right away, figuring out a realistic project or practice to take on makes sense. Or, they might feel comforted by developing a plan for environmental action during the summer months (e.g., a conservation project or advocacy work). Teaching kids the tools needed for making life decisions is an invaluable life skill.

Cocreate. Collaboration is the key in supporting your child's vision because it combats isolation and overwhelm. However, you don't want to lead or get ahead of your child, or they'll find an excuse to bounce. Try to stay lateral, like an ally. Imagine that you're walking alongside your child in the darkness holding up a lantern to help them see better. In the script for vision, I'll describe a mapping process which can be done collaboratively but that could also be a vision board, a journaling exercise, or an improv. Some of my favorite creative prompts to use in the cocreative process include:

- Draw a picture of a bridge showing where you are now on one side and where you want to be on the other
- Draw the crossroads where you are and the different paths

- Draw a map of the different hats that you are currently wearing, and feel free to include some that you are considering to wear
- Map all of your current roles, identities, and responsibilities

Ultimately, you're supporting your child in figuring out how to stay true and build on their values, hopes, and dreams. Concretizing steps in a clear way by activities such as writing down a step-by-step plan of action can be useful. The editing process—what you say no to or discontinue—is part of your child envisioning their future. Creating an accountability plan with your child, whether it is through a buddy system, calendar reminders, or a climate date, can be supportive.

Weave in intergenerational work. As you support your child's vision, one way to increase their sense of empowerment is by partnering with adults in some fashion (and that might include you). Your child might want to write a letter to state representatives, attend marches, or participate in ecological restoration projects. You can offer to be an active participant in their vision. Sure, they might not want you at a rally with them when they're with their peers, but they might say yes to working on a poster or sign together. Even if you think they'll hate the idea of parent-child collaboration, it's still important to offer it so they know it's on the table. Plus, you can help connect them to other less embarrassing adults as well (meaning, literally, anyone but their own parents).

Now, for a vision script to help you understand the idea. This is a longer script, which shows how a parent might support emotional processing as well as develop vision.

CHILD: I'm really stressed out. I've got too much on my plate. I have finals next week, my soccer tournament is this weekend, and Ava's mad at me again . . . and I had another nightmare last night about a wildfire burning down our home. It's just too much. [*Cries a little*]

PARENT: Oh, honey, you've so much going on right now. It's a lot.
[*Pauses, maybe places arm around her child's shoulders*]

Child: You can say that again! I hate my life.

Parent: Well, you named a bunch of things that are on your mind. What if we just focus on one to start with?

Child: Ugh, okay. My nightmare. It was scary.

Parent: Sure, can you tell me about it?

Child: There was a huge wildfire and . . . [*Describes details*]

Parent: Got it. That does sound scary. Sometimes when we're so busy we don't have space to unpack our emotions. So our emotions come out in our dreams instead. I'm wondering if the recent wildfire might have affected you more than you realize.

Child: Uh-huh. [*Some tears*]

Parent: It's normal to be scared of wildfires—and about climate change. But it doesn't help you—or anyone—to think about it all of the time, either. Something that might help—and this might sound odd—is scheduling in a time to check in about your climate emotions each week.

Child: Huh? What are you talking about?

Parent: Well, nightmares are no fun. So turning toward what feels big and scary during the daytime can actually make your worries shrink because you're facing them. Does that make sense?

Child: Yes, but I don't really want to think about it.

Parent: It's not an easy thing to do. What if we do an exercise together that might help?

Child: [*Sighs*] I guess.

[*PARENT GRABS SOME BASIC ART MATERIALS, LIKE MARKERS AND PAPER*]

Parent: Okay. Draw a circle in the middle of the paper and fill it with colors, lines, and shapes that show how you are feeling. . . . [*Pauses until done*] Now draw a larger circle around that . . . include what helps you to feel better.

Child: Okay, I can do that. [*Draws*]

Parent: [*Sits quietly, or comes back when child is ready*] Now, let's look at this together. How was it to do this exercise?

Child: Good. It helped me to put all of my thoughts and feelings down.
Parent: Glad to hear it. Do you want to share anything about it?
Child: Well, that's a big wildfire burning down our home . . . I drew rays out of it like a sun with some things that might help—like not checking my social media before bed or joining Fridays for Future. I put a ring of blue water around the rays to contain the spread of wildfire and my worries.
Parent: Wow, thanks for sharing. I can see a lot of dark colors where the fire is, how out of control it is. I also see those rays reaching out for help. Which ray is easiest to start with?
Child: I want to join the school club because I can make friends who will understand what I'm going through.
Parent: Makes sense. Do you want to look into that this week?
Child: Yes, I'll ask my friend who's in it.
Parent: I'm here anytime. Anytime you feel stressed or lost, you can try mapping it out like this and see if it helps.

Vision is simply having the willingness to explore your child's feelings and ideas nonjudgmentally with them. You can share what you notice about their artwork as long you refrain from judgments and use it as a springboard for questions. When in doubt, just stay with the metaphors in the art. I use versions of this exercise all the time with clients. When they're overwhelmed, transferring that overwhelm onto paper helps by (1) externalizing their emotions onto a physical separate object, (2) growing insight and perspective, (3) stimulating their ability to problem-solve or think creatively, and (4) communicating thoughts and feelings to another person. I've had clients examine their peer or romantic relationships, their college and gap year options, their professional development, or work-life balance using this tool.

The three V's are a humble reminder that our kids are on their own life trajectories. We're not there to control them but rather to support them in gaining a sense of autonomy and independence that will serve them throughout their lives.

4. The Framing: Nurture an anti-doomer outlook. The news is full of positive climate stories if we look for them, and yet it's the negative news that dominates. This is partly due to what our mainstream media outlets choose to cover and know will sell, but it's also due to our inherent *negativity bias*. Our inbred negativity bias zeros in on threats and potential problems because we are, after all, wired for basic survival. While it makes good sense to scan our surroundings for safety, it doesn't make sense to transmit fear and negativity straight to our kids' hearts.

Emotions are contagious. If parents walk around in a perpetual state of gloom and doom, then (surprise, surprise) we'll raise a generation of doomers. Not only will we pass along an inheritance of a marred Earth, we'll also be passing along a lack of agency to do anything about it!

As tough as it can feel, it's essential to instill rational, active hope and trust in our kids. At the end of the day, kids need to feel that adults are trusted allies—which is, by the way, exactly what we should be!

Here are few preliminary ideas about how to nurture an anti-doomer (there will be more on this in chapter 10).

Acknowledge progress in the right direction. From restoration of critical watersheds and electrifying school buses to community greening efforts and youth speaking truth to power, progress and evidence of a growing culture of care is everywhere. Carve aside a moment each week or month to share positive stories, news clips, or reels as a family. Family members can rotate this task or each family member can share something each time. Digesting positive stories replenishes and regulates us, steering us away from the rails of doomerism.

Balance truth-telling with empowerment points. In her book *How to Talk to Your Kids About Climate Change: Turning Angst into Action*, author Harriet Shugarman defines "truth-telling" as sharing with kids the realities of human-caused climate change. But she also recognizes that parents must find ways to do so without burdening kids with too many distressing details. She recommends pairing five empowerment points for each difficult truth shared (a 5:1 ratio) so that kids don't end up flatlined.

Raising Anti-Doomers

If you say to your child, "The coral reefs are dying out because of climate change," don't just stop there and leave them to deal with it! Don't be an Eeyore and create a mini-me Eeyore, or force kids to be a Tigger to help bring your spirits up. *Your* job is to keep conversations fluid, and periodically refuel with positive news. As an example, five empowerment points here might be: (1) coral reefs can be restored naturally; (2) humans have developed technologies to restore them (e.g., biorock); (3) the United Nations leads regional programs to protect coral reefs; (4) there are many nonprofit organizations focused on this issue; and (5) we can visit aquariums to learn more about them. Offering empowerment points to balance truth-telling can help build an anti-doomer mindset early on!

Language is clutch. Words are powerful. We can't underestimate how they elicit emotions. In the climate space, there are many different terms, and it's important to consider the visceral effects of the language that you choose.

For example, I might use "collapse" or "chaos" in this adult book and in sessions with adult clients. However, you won't catch me using those words in front of my elementary-school-age daughters or with children. Those words are heavy hitting; they ring with a sense of finality, devastation, and despair. Instead, I use the terms "climate change," "changing climate," or "climate crisis," or I might just describe what's happening (e.g., "Earth's getting hotter," "pollution is hurting Mother Earth)." With teens, however, I might adopt their own language to validate their experience.

In chapter 5, I mentioned climate empathy, a person's care and concern about earthly life. This term sheds a more positive light on a person's experience of climate distress rather than pathologize it. As another example of the potency of language, eco-philosopher Glen Albrecht has suggested that instead of defining our age as the "anthropocene" (meaning the age of human destruction), we call it the "symbiocene" (meaning that we are transitioning to an age of interdependence). This difference is not mere semantics; language affects our mood and our perspective. Anthropocene elicits a chill, while symbiocene elicits hope. This also applies to parenting:

instead of saying "do not do X" to our child, can we instead try saying, "can you please do X?"

In my practice, I often hear kids say things like "I have anxiety." There's something strange in hearing an eight-year-old describe themselves this way. It suggests that they are thinking *There's something wrong with me*. Meanwhile, I'm thinking, *What's wrong with our culture that a child is not only feeling such a high level of anxiety but taking on this symptom as part of their identity?*

The word anxiety, although normalized at this point, is also rigid, stigmatizing, and limiting. But emotions come and go; they don't define us. Existential therapists believe that confronting anxiety is a normal part of life. There's a risk that kids, who are often obsessed with identity and what it means for their social relationships, might over identify with this emotion, internalize it as a part of themselves, and create all sorts of stories around it. In fact, it can simply be an emotion that is resonant for them during particular moments of their lives. Shifting language slightly to "I am *feeling anxious* about—" is more flexible and forgiving.

In a 2024 *New York Times* article, "Are We Talking Too Much About Mental Health?" research psychologists at the University of Oxford expressed concern that self-labeling or self-diagnosing—fueled by TikTok and the web—can also be detrimental for young people. The article highlights research in support of "targeted, light-touch interventions," which can be effective in decreasing anxiety in younger children. There is a "less is more" takeaway here, particularly of note for parents who tend to be overinvolved or helicopter-y. In my practice, I'm always trying to figure out what balance or combo of therapeutic interventions works for kids. In some instances I've found that too much emphasis on thoughts and feelings can backfire and make a client feel worse. Guided meditations, validation, and relational support can be more effective for some clients than targeted cognitive behavioral therapy interventions. For a tween client I met with experiencing obsessive-compulsive symptoms, rather than use "OCD," I supported her in coming up with her own name to categorize behaviors that she feels she must do again and again.

The important takeaway here, from my perspective, is that we don't want to overdo talk and worry about mental health at too early an age because it can lead to over-rumination and over-identification with mental illness (e.g., "it's my ADHD"). And yet we also don't want to cause shame or confusion by not talking about it. Treading lightly is wise while using language that is non-stigmatizing ("everyone's mind works differently and has different strengths and challenges"). I think we can lean on strength-based frames that center strengths, abilities, dreams, and goals above fear- and anxiety-based narratives.

Since we're on the topic of language, I will now make a plug for the term *anxcitement*, an emotional hybrid of anxiety and excitement. It's not uncommon to have difficulty distinguishing between anxiety and excitement because the sensations associated with them are similar (butterflies in stomach, rapid heartbeat, restless energy). I use this term all the time with kids because it offers a flexible, healthy, and normalizing way to communicate a particular experience. Exploring language in this way is a good reminder that emotions are nebulous and complex. As a result, our relationship to them shouldn't be so rigid.

Before moving on, I'm going to take a stab at some of the questions that might be on your mind, Dear-Abby–style.

What Do I Do If . . . ?

My child is vocalizing their concerns a LOT and I'm worried about how this might influence their siblings?

Answer: Contain and redirect! Is there a way to carve out some one-on-one time to address your concerns, but not always address them in the milieu? When kids feel heard, validated, and supported, it can help relieve some of their distress and need to vocalize. Also, helping your child channel distress into external actions can feel relieving. Distress or doomerism can be contagious, so it's important to help prevent spread.

My child isn't vocalizing any concerns, but I know this stuff is on their mind a lot.

How to Talk to Your Kids Amid a Brewing Existential Sh*tstorm

Answer: "Honey, I'm your mom/dad, and I can tell that you're worried about something. Do you want to talk about it? Draw a picture or write a poem about it?" For younger kids: "Do you want to draw a picture of your worry monster?" It can be helpful to explain that worries actually get bigger the more you keep them locked up inside. Sharing them with others helps make them smaller.

I'm about to bring up this ginormous topic for the very first time. How should I go about it?

Answer: First, take a PCP. Breathe, walk, give yourself a hug, self-appreciation—whatever feels right to you. Now, take the pressure off yourself. Remember that you are offering your child an invitation or bridge to connect, and you can't control the outcome. Say goodbye to those perfectionist demons.

I'm worried that I will be too emotional.

Answer: Again, start out with a huge whopping portion of self-compassion. You're a loving and caring parent who also has some big feelings about tough real-world issues. Remember: It's okay to cry and allow feelings to naturally bubble up. In fact, it's healthy modeling for your kids when you meet your feelings with compassion and they witness that. If your emotions are feeling so intense that you start to become dysregulated, or out of control, then state the obvious: "Mama has big feelings about this, but it's important that we check in from time to time. Let me take a minute to regroup and we'll try this again later."

What if your political beliefs are different from your kids?

Answer: Healthy debate and disagreements within families teach kids about tolerance, multi-perspectives, and how to be respectful across political divides in the world. Silence and secrecy foster shame. If you're worried and your kids are not, then it's still fair to mention *your* thoughts and feelings. On the flip side, if your kids are worried, and you are less so, just know that you're holding space for what's important to them and offering that invitation to come to you while also teaching about tolerance and respect.

What if you've brought up the topic before and it didn't go well?

Answer: All good—just try, try again! No need to attach to the results. Just review some of the basics in this chapter, read the cues, listen, validate, and consider this all part of learning on the job (that's parenting for you).

Preparing Kids with an Essential Life Skill: How to Navigate Difficult Conversations

It's one thing for parents to take on difficult conversations with kids. But what about when kids go forth and face difficult conversations themselves about challenging issues?

Kids can learn from peer mentoring about this. You can encourage them to listen to youth who are already out there—the ones bringing on badass lawsuits, protesting or campaigning, speaking to elected representatives, and living in alignment with their values. However, because this is a book by a parent for parents, I'll share with you a couple of brief thoughts about this.

1. *Come from empathy.* Social psychologist Jonathan Haidt wisely said, "Empathy is an antidote to righteousness."[1] In this world, in this time, the healing we need to see will come from building bridges toward one another and by softening these spans of connection. Coming from a place of empathy is paramount. It's all too easy to hop on the slash-and-burn hater speak train and hurl out judgments into cyberspace. But it's incredibly damaging and dismantles are attempts to come together. Teaching our kids about empathy starts in our homes. Assuming the best in people as you and your family go about their days is a start.

How to Talk to Your Kids Amid a Brewing Existential Sh*tstorm

Take a moment to go inward. If it feels right, then close your eyes, and deepen and elongate your breath. Imagine where inside of your body empathy dwells. Place a hand there. Notice: What sensations are present? Does the place have a particular color? As you breathe in, feel the connection between your hand and this place where empathy resides. After a few minutes of connection, take this contact point with you, remember the color, and maybe even create empathy art about this. Return to this place in your body (or your artwork) before having difficult conversations. (And, of course, teach your kids about this too!)

When having difficult conversations with others, your goal isn't to convince them of anything, but rather to cocreate a respectful platform for present and future discussions. Since communication is a two-way street, you can only control your side of this. If it goes sideways, remind your kids that they're modeling something new in how they approach these conversations, and just their attempt is meaningful.

2. *Revive the lost art of listening.* I love the saying "listen to understand, not to respond." (I'm pretty sure I heard a contestant on *The Bachelor* say that, but that's beside the point.) Fundamentally, it's the only way to deeply listen and connect with another person. Your own nagging agenda must be put aside to fully listen. When you listen, you show patience and goodwill. Rather than hatch a counterargument, you spend time ensuring that you are getting their point of view. Whatever you do, don't take the bait of debate—it will only lead to both sides aiming to win and neither side properly hearing each other. When I think of what it means to be a good listener, I always recall one of my clinical

supervisors. She'd sit quietly listening to me go a mile a minute, watching me with gentle, patient eyes, nodding in attunement, and saying "mmm" a lot. I felt so incredibly met and held by her. As I vow to work on my listening skills as a parent myself, I tap into this memory sometimes—she's a helpful guide. For extra reading that will help to elevate your listening skills and ability to connect with others, check out meditation teacher Oren Jay Sofer's book *Say What You Mean: A Mindful Approach to Nonviolent Communication*.

3. *Hone your nonviolent communication skills.* Nonviolent communication (NVC) was developed by Marshall Rosenberg in the 1960s and 1970s as a communication tool to improve understanding and connection through empathy. A few of its golden nuggets are connecting empathically with what's alive in the other person, expressing your feelings and identifying your needs, and making clear requests without forcing the matter. Its four basic components are: Observations, Feelings, Needs, and Requests.

On the NVC Academy website (https://nvcacademy.com/media/NVCA/learning-tools/NVCA-feelings-needs.pdf), there's a list of universal human needs and a list of feelings (emotions). I'll often ask clients to attune to their current needs and feelings by circling them on this list. In addition to teaching these foundational skills to kids, it can be extremely helpful for parents to shore up their relationships with partners/coparents as well. Sharing the results creates mutual vulnerability and a chance to practice empathy. It's when we ignore or fail to recognize our deepest needs that we resort to manipulation, coercion, or other unhealthy means to get them met. Self-awareness and empathy will help kids to keep difficult conversations from devolving into screaming matches.

My hope is that you're now feeling more ready than ever to broach conversations with your kids about tough real-world topics. Remember: You don't need to have all of the answers. Just be gentle with yourself, be flexible and willing to go with the flow. At the end of the day, those of us in the Parent Club are used to stumbling around in the dark, figuring things out on the fly, and making messes and cleaning up. You've got this!

The Creative Exercise that follows is a fun family art directive that will generate plenty of fertile ground for discussion and connection.

CREATIVE EXERCISE

ANTI-DOOMER PHOTO PROJECT

Materials:
- Camera
- Photo album or book

or
- DIY book with cardstock, paper, cardboard, photo album
- Glue, binding materials (e.g., duct tape, hole punch, cording), stickers, decorative papers or tapes
- Optional: Pinterest or other virtual photo board

Setup:
You can do this alone or with family members. If doing this as a family, decide on a time frame or deadline for the project (anywhere from one week to three months is good).

Everyone will take individual photos during that time. Consider keeping a separate journal to track your thoughts and feelings along the way.

When you are satisfied, go ahead and develop your photos and create a photo book. Edit it as desired. Consider titling book and writing captions for the photos. Note: This can be done virtually, but creating a tangible book is highly encouraged.

Prompt:
As you go about your life, take photos that inspire you, fascinate or interest you, or give you hope. Your primary focus should be the natural world and any resources that inspire wellness, but you can include other photos, too.

Process:
As you take photos and create the book, notice without judgment any feelings, thoughts, and sensations arising in your body. Consider recording these emotions so that you can pair them with the photos at the end.

Product:
Set aside a designated time to share these books together as a family. Be mindful about refraining from judgments (good or bad). Consider setting ground rules for sharing these books in order to create positive space.

Reflection questions:
- What did you enjoy about the process? Or not enjoy?
- How do you feel in viewing your book? What feelings, thoughts, and sensations arise?

Specific questions for family projects
- Were there any common themes between all of the books?
- Did you find the photos in other family member's books inspiring?
- Are you interested in integrating the books somehow or creating a collaborative work?

EIGHT

Growing Your Identity

Practices for Staying Engaged

"I sensed in my feelings of despair and outrage, there was power in that, a mother's love for her children."

—Kelsey Wirth, cofounder of Mother's Out Front[1]

When I turned forty, I found myself reckoning with my age, surging with an identity struggle much in the way I did in my early teen years and in motherhood. *Who was I?* But more pressing at that time: *How could I possibly be forty years old?* While grateful to have my heart full and for my circumstances that had permitted so many years of vibrant life, I was also bowled over by genuine disbelief and a budding, nagging curiosity about what was next.

In that midlife crisis sort of way, I sought to redefine myself with random, grand plans: *I'll sign up for aikido class . . . I'll take belly dancing classes . . . I'll finally get a tattoo.* In the end, I threw a big birthday bash. I squeezed into a Spanx for the first time and my circa-1997 navy blue sequined junior prom dress, and then gyrated on the dance floor with other folks my age to a curated dance mix of early 1990s music.

But, best of all, was the open mic. I invited folks to share embarrassing entries from their old childhood journals. (Full disclosure: I got this idea from a San Francisco-spun event called *Mortified*, in which people take the mic and share hilarious journal entries in front of an audience.)

Why recall this here? Well, I've always had a personal affinity to child-adult integration work. And I see massive potential in how it forges new possibilities in parenthood as well as in adulthood. I do such integrative work in my office, supporting adult clients in reconnecting with their child selves to aid their personal healing and growth. Whether it's a parent's difficulty in validating a child's feelings, a distant/fraught relationship with a family member, a traumatic memory, or lack of confidence, sometimes meeting your inner child helps.

For many of us, to better understand and help our kids, we might revisit our kid selves. Doing so can increase our empathy and understanding of our children and decrease reactivity that is more to do with our own childhood grievances and traumas.

To that point, this chapter is all about identity and new growth, with a particular focus on the parent work that will improve our ability to relate to, understand, and support our kids. There's a Gestalt-inspired exercise for you to bring your adult and child selves into the room for a long overdue conversation, some reflections on the awkward ripeness of puberty, and the concept of identity sculpting, an active, conscious process that can be tapped into at any age that I will introduce. Forging active (rather than passive) identities ensures long-term engagement and commitment. Repeat after me three times: *Parent passivity must become passé. Parent passivity must become passé. Parent passivity must become passé.* Now let's begin!

Growing Your Identity

Getting Acquainted with Your Child Self

To show up authentically engaged, and to help us in parenting, it's important to develop a working relationship with our wound baggage.

In the Parent Pause that follows, I offer extended time for you to reconnect with your child self, wherever it may lead you. Some of you might be more (or less) excited about this. However, this work will benefit your child by refreshing your memories about what it was like to see the world through a young person's eyes. It will also benefit you by creating a through line of identity—an integration of your child and adult selves. This will help you to cultivate a strong sense of self as you journey deeper into parenting during a polycrisis. I caution you not to skip this.

Please note: If this exercise makes you feel uncomfortable or you're worried that you might be triggered, you're not alone. If you have significant childhood trauma or concerns, then I suggest doing this exercise with a professional or highly trusted adult. If you do it alone, I'd recommend—more than any other exercise in this book—debriefing at the end with somebody about the whole experience and what came up for you.

A HEART-TO-HEART CONVERSATION WITH YOUR YOUTH SELF

Part I: Setting the scene (staging)

Spend at least thirty minutes (or more) revisiting your adolescent self. I recommend an hour or two, or even the span of a weekend. Consider playing music or songs from that era to help evoke the spirit of the time. If you have access to photo albums, pictures, yearbooks, schoolwork, art, journals, or memorabilia, get these out and spend some time combing through them.

Some reflection questions as you ponder your adolescent self (and journaling is always welcome):

- Who were you in your teenage years? What roles did you have?
- How did you see yourself then?
- How did others see you?
- What did you stand for?
- What attitudes did you hold about the world?
- What memories do you have of your relationship with the MHW?
- What child "parts," unfinished business, or wounds are still with you as an adult?
- What happens if you approach these parts now with love and compassion?
- What needs to be resolved to move forward with your parent identity?

Part II: Authentic role-play

For this part, choose between an experiential (option A) or a written (option B) exercise.

Option A uses the Empty Chair Technique (from Gestalt therapy). Set two chairs facing each other and enact an imaginary conversation between your adult self and your youth self. Picture yourself at an age that feels resonant or that you remember well. Feel free to use props, like two different hats or items to distinguish the two versions of you. You can assume what might be very different postures/mannerisms in these two roles. You'll move back and forth between the two chairs as you create a spontaneous dialogue with your adult self in one chair and your youth self in the other one.

If you'd like a bit more structure, then try digging into a specific topic like climate change. As a young person, how did you feel about nature/the environment/global warming? What did you notice? As an adult, how you feel about it now? What question do you want to ask your youth self about it?

Or what would your youth self say if they were alive today? Can you move toward a sense of resolution? Consider recording this conversation for future reference.

Option B approaches the same exercise in written form. Select two different colored pens or markers to distinguish youth and adult voices, and write a dialogue imagining your adult self meeting face-to-face with your youth self.

Post-exercise reflection questions:
- Which self started the discussion?
- Was it hard to listen or to speak in either role?
- What emotions or physical sensations came up in each role? (Flip back to chapter 5 to reference the Somatic Sensations List and Climate Emotions Wheel.)
- What similarities or differences do you perceive between your youth and adult perspectives?
- What did this exercise teach you about your parent identity? Can this imaginary meeting guide you in some way?

Be sure to take ample time for reflection, journaling, and sharing afterward. If you feel triggered, please go to the support resources in Appendix C.

I think of this exercise as threading the needle between the wisdom of your younger years and the wisdom that you've accrued since, and how this integration might better serve your parenting. My hope for you in having led you into this tender nexus is that you feel something differently, a new insight flashes, shifts your perspective, or tends to a wound that will serve you as you venture forth as an engaged parent. Whatever it is, be sure to seek the support that you need, and show up with a deep, deep well of compassion, which offers vulnerability a soft place to land.

As we seek to integrate ourselves more fully with—and possibly heal—parts of our child selves, we prime ourselves to better show up for and support our kids. We prime ourselves to better understand their points of view, to better empathize with their situations, and to better share in their passion and sense of urgency. Most importantly, we prime ourselves for reinvention in order to take the bold and meaningful action required to meet this moment.

Remember the Awkwardness of Puberty? Well, Time to Revisit It!

Psychoanalyst Erik Erikson identified the key struggle of adolescence as forming an identity.[2] If you don't figure out who you are, what's important to you, what you stand for, then you might slip into despair. During this period of life, it's pretty typical to toggle between a sense of who you are, who you want to be, and the actual constraints that affect these desires. This confusing brew is akin to a pinball machine, where you bounce around the socio-cultural landscape encountering various entry points and obstacles, your hands at the joystick. For better or worse, in today's world identity extends into the far reaches of cyberspace as well.

Identity development is a critical part of adolescence: Young people are constantly trying to figure out—and assert—who they are through their actions, behaviors, emotions, and convictions. That's why it can be so intense and dramatic to witness this coming of age as a parent. Sometimes teens are not only trying to convince others of who they are, but they are also trying to convince themselves. Developmentally speaking, youth can be idealistic, action-oriented, willing to take risks, passionate, opinionated, future-focused, and less encumbered by responsibilities because many are cared for and supported by their families.

Puberty is one giant, unfolding, excruciatingly awkward metamorphosis. A kid's body is secreting new hormones, their biochemistry is out of whack, periods start abruptly, pimples pop up inconveniently, growth spurts happen unexpectedly, sexual urges come online, and so on. The unpredictability and sheer lack of control over growth, experienced as if on IMAX-level display, can be uncomfortable at best, mortifying at worst. (For me, as with a whole slew of Gen X/elder millennial parents, Judy Blume books like *Are You There, God? It's Me, Margaret* just about saved my life.)

Now let's link all of this puberty-talk to parenthood. Parenthood is a time of major identity transformation. But because there's all of this "raising a new human" stuff happening, this life shift doesn't receive as much attention as it should. Anthropological terms "matrescence" and "patrescence" connote this transition into motherhood and fatherhood and actually parallel some aspects of puberty, adjusting to bodily hormonal changes and relational changes. Self-consciousness and vulnerability bubble up once again, particularly as you might feel pressure to project an acceptable—or even admirable—image to the rest of the world. But the supermom and the popular kid are often victims of their own identities; in other words, they are slaves to proving themselves over and over. Just as puberty requires kids to continue on despite all of the big changes that are happening while peers react, parents must also do the same.

My mom has always said that the older you get, the less you care about what others think. As usual, she was correct. And if you're lucky, there comes a point in parenting when you start to feel some degree of confidence or authenticity. Other ways to describe this new stage are "you stop giving a shit" or "you let it all hang out." Perhaps the most obvious manifestation of this in my life is how comfortably I walk the morning schoolyard in uncool exercise wear and nary a glance in a mirror before interacting with the horde of parents dropping off their kids. Honestly, I've almost felt as sassy as Beyoncé on the catwalk as I skip down the path to the parking lot after dropping them off (key words: *after dropping off*), the words echoing in my mind: "I don't give a shiiiit." (And that's about 95 percent true!)

PARENT PAUSE

Can you think of a moment of parent-borne confidence that you viscerally experienced? Go ahead, feel it in your body, your posture, your energy. Now, own it, own it . . . Right on. Stay there for as long as you'd like. Throughout the day, reconnect with this source of strength, and throughout the week, year, etc., as well. Let's call this the Parent Powerhouse Pose. (This exercise is a cousin of the Shape-Shifting the Dialectic exercise in chapter 1.)

When preparing yourself to embrace change, you may encounter feelings of shame, embarrassment, discomfort, internal resistance, avoidance, numbing, self-consciousness, and defeat. For example, in developing a climate identity, you might voice your values at work, bring the topic up at a party, make a community presentation despite feeling shy, or write an op-ed for the local paper—all behaviors and actions that can really push you out of your popcorn-and-PJs comfort zone. As adults tell kids all of the time, growth stretches you in new ways.

As you sculpt your identity, a growth mindset is your friend. Self-compassion is a must. Fear not: You can make adjustments along the way! If things feel intense, then help yourself or your kids to slow down, seek support, or scale back. If you don't feel you are doing enough, then step up a notch. Just be patient as old blocks, defenses, and habits present themselves (and spice up the journey). Not only will conceptualizing an identity help to focus, ground, and buoy you, it will also be relieving for kids to watch adults assume the responsibilities that are really theirs in the first place. As you push yourself and stretch yourself taut like a rubber band (but not too taut!), please continue with your PCPs. (In fact, please do one right now!)

Identity Sculpting in Action

Identity is a mingling of experiences, memories, relationships, culture, gender, sexuality, ethnicity, politics, career, class, religion, values, salt, and pepper (okay, not the last two . . . just checking to see if you're still with me). Identity gives you a sense of belonging, even *drishti*, which bolsters confidence and well-being. Identity is also a way to anchor yourself in your values and hopes, no matter what crazy stuff is playing out in the world. A strong identity is like a sturdy platform that you can stand on—or come back to—as you set forth on an adventure.

The term "identity sculpting" is an apt metaphor because we can be sculpted like lumps of clay, being shape-shifted and transformed through the creative process. You can add or subtract according to your liking—a bit more activism, a bit less anger. Sculpting is an active not a passive process. Rather than just lie about baking in the sun, we have some real agency over how our lumps are molded in this world.

With conscious effort and intention, we have the ability to shape our identities. How awesome is that? Sure, this might sound like a major overhaul, but do not fear! All it really means is your willingness to build off your many strengths and talents, and a willingness to grow in new ways, even as you might feel vulnerable at times while contorting into new shapes.

The concepts that follow can be shared with kids in a developmentally appropriate manner. That might mean simplifying to one main point, such as "taking action helps you feel better." But this chapter is more indirectly in service of your kids, to push you in new ways. Now, let's get more concrete with illustrations, exercises, and stories. We'll start by zeroing in on the relationship between our emotions and actions.

Visual #1: The Climate Emotion-Action DNA Model

Imagine a DNA-like double helix with action and emotion strands continuously interweaving; the strands must be integrated for holistic integrity. If they are braided in this way, then channeling emotions into action can

THE CLIMATE EMOTION–ACTION DNA MODEL

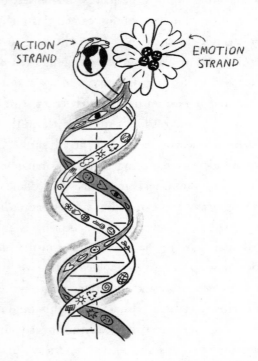

feel relieving, healthy, and productive while also acknowledging feelings that surface *through* action that helps to sustain resilience.

This concept builds off the emotional metabolizing you did in chapter 5 (e.g., the climate emotions, the pellet, the compost). Bring that rawness and heat here and take it to the next level.

This is similar to the yin and yang energies in Chinese philosophy, which hold that opposing forces are complementary. Examples of yin and yang permeate and abound in nature: Mother Earth (yin) receives the light and warmth of the sun's rays (yang). These forces are life-sustaining, and so it is with emotions and actions as well. Emotions are internal responses to life that find relief through external actions. Metabolizing your emotions is a form of receptive surrendering (yin), while taking action is a form of energetic manifestation (yang). When you tap into a flow between these states, it can function just like breathing: breathe out emotions, breathe in action, and repeat. This synchronicity is healthy, and one without the other can lead to imbalance.

Anger is often considered the most productive emotion because it leads people to take action more so than other emotions. Several studies indicate that anger in particular is associated with greater engagement in pro-environmental activism and behaviors, predicting better mental health outcomes than other climate emotions. This supports the rationale for the Climate Emotion-Action DNA Model.

The tough emotions that we experience toward any injustice—anger, fear, betrayal—can be leveraged for action. Internal emotions find sturdy relief through external actions. Several studies point to how environmental activism reduces climate-related distress. Psychologist Susan Moser has discussed that the relationship between climate activism and climate emotions is bidirectional, meaning that high levels of climate distress motivate action, while engaging in climate activism relieves distress.[3] But we have to keep in mind that there's a tipping point with anxiety. It can be an adaptive response to a real-world problem, but if it overwhelms a person's capacity to cope, then it becomes self-defeating.

Managing anxiety is critical so that it does not become paralyzing. Action requires periodic rest and rebalancing to avoid chronic stress, fatigue, and over-activation. It is common for activists—and those adjacent—to approach burnout from time to time. Reactivity stems from an activated stress response. It can burn hot and fast and work against our health—and even against the work that we are trying to do. As tempting as it might be to go at the causes we care about at full-throttle, we must pace ourselves or plan high-intensity interval engagement keeping in mind our well-being. Our kids are good reminders that we are training for a marathon, not a sprint.

Staying attuned to our fluctuating emotions and instituting consistent self-care practices are sensible measures to mitigate burnout and exhaustion. While both under- and over-involvement in an urgent crisis are problematic, we have to try to position ourselves to the best of our abilities. If we feel like we can do more, we take a step forward. If we feel like we are on the edge of burnout, we step to the side and recoup. Call it "the Climate Dance," if you will; there is both a sense of leading and following at different times. Many therapists support climate activists, scientists, and other climate-involved folks with burnout (even climate psychologists).

Visual #2: Climate Action Venn Diagram

The next illustration to usher along your identity sculpting is the Climate Action Venn Diagram, a visual designed by Ayana Elizabeth Johnson to help us get organized around the action that we are willing to take. I love exercises that shift problem-saturated narratives to solution-focused narratives. It's a mindset shift that can energize, direct, and create new perspective. Check out the Venn diagram and then you can get started on the exercise below to help you and your kids get crackin' with your

Credit: Ayana Elizabeth Johnson. Reprinted with permission.

Growing Your Identity

identities. You can help your kids along the way. Feel free to share as you go or at the end.

Grab paper, pencil, eraser, and maybe some markers or gel pens. Consider taking this exercise outside in some way, contemplating it while on a walk or sitting in a quiet setting in the MHW. What natural setting or element might support you here, or do you feel intuitively drawn to in this moment (e.g., a meadow, forest, water feature)? If you're able to go there, go!

Process

Draw three overlapping circles (these can either be drawn freehand or you can trace a round object like a plate or bowl). As per the Venn diagram, all circles should overlap in the middle in what I'll refer to as the "union."

Label each circle with a question: (1) What brings me joy? (2) What am I good at? (3) What work needs to be done?

Next, answer those three questions within their respective circles.

When finished, reflect on what you notice. If you see patterns or places with potential overlap, please write those ideas in the union. Or you can work on the unions as a family.

When your process feels complete, you can color in the union or add any desired embellishments. You're welcome to create a final version of this exercise using ink or markers. Alternatively, you could create a larger image of just the union to showcase your identity.

Write in your journal and/or share your Venn diagram with an accountability partner, family, or group, being mindful of any thoughts or feelings that come up.

Reflection questions:
- What has been revealed through this process?
- Does it feel sustainable?
- Can you imagine taking next steps?
- What obstacles might arise, and how might you work around these?

Parents can exercise their voices and actions in so many ways, whether speaking up at a PTA or school board meeting, initiating green ideas in their workplaces, or forming a community group focused on sustainable living. Yes, it might feel embarrassing or scary to take these steps at first, but ultimately the goal is to lean on your skill set and create a role that feels natural, manageable and, ideally, interesting. Remember how it felt when you did something new for the first time—whether riding a bicycle, standing up in front of the class for show-and-tell, or facing an ex? When we do something new, we put ourselves out there, and it can feel like all of the vulnerables. If you remember, parenting and climate change are two vulnerable peas in a pod!

E-piphanies or Aha! Moments

In the Parent Identity Spiral in chapter 1, you might have seen the words "e-piphany!" or "aha! moments." For many folks, identity sculpting does not come out of nowhere. Often there's a discernible moment or event that stirs a deeper reckoning. This reckoning can be the impetus toward change.

British psychotherapist and founder of Carbon Conversations, Rosemary Randall, calls these inflection points—at least those related to waking up to the environmental and climate crises—"e-piphanies" (i.e., ecological epiphanies). An e-piphany is an emotional moment that changes everything for you, a moment when you jolt awake with a deep recognition of what is happening to the environment and your vision narrows to a more decisive focus. It is the moment that you feel deeply and cease intellectualizing. As Randall said in a podcast interview, it's the moment you say to yourself, "Oh wow, this is me . . . this is my life . . . this is my future . . . this is going to happen in my life time."[4] For the Parent Club, an e-piphany might center more on their child's future than their own.

Many environmental activists have an e-piphany story, something that broke through their MA and cracked them open to a new way of seeing things with urgency and commitment. Indeed, I've talked to and heard from many folks who describe it as an awakening. Following this e-piphany is a snowball effect: their motivation increases, then their actions and behaviors, and then overall engagement and possibly a new worldview.

Since the ideas in this book are broader than climate change, I've included aha! moments as an alternative to e-piphanies. For many parents, aha! moments spark when an issue becomes personal, threatening their family life in a real way. This makes good sense, based on what we've discussed about humans prioritizing immediate concerns. Suddenly, an issue is not just some abstract matter taking place hundreds of miles away, but rather a local issue piercing into the emotional heart space of their communities.

For others, the aha! moment is more of a gradual realization that springs out of a felt sense of parental dissonance between hopes and tough realities. For still others, these moments flow from a rekindling and reconnecting with deep core values through parenthood, such as a love of the MHW. Whether in a nanosecond or over the course of many years, a sudden jolt of realization illuminates loss and fear, eliciting clarity.

My own climate awakening felt like an apocalyptic version of the Rip Van Winkle tale. The world I woke up to in my late thirties, after seven or so years of disavowal, featured the blazing wildfires in California. In the fall of 2017, a few months after the birth of my second daughter, my husband and I decided to leave town with our family to escape the deteriorating air quality due to wildfire smoke.

Already at that point wildfires had moved from the something-that-happened-elsewhere category to the something-that-was-affecting-us category. As a family able to avoid health exposure to smoke and stay with our family friends upstate, we were lucky. (Many folks did not, and do not, have the resources to escape; they simply have to stay put, and grin and bear it. Still others may not grasp the associated risks, whether or not they have the means to stay or go. Smoke, of course, is just the tip of the iceberg. Over the years, people I know or were connected to have experienced evacuations and lost their homes altogether.)

But in the fall of 2019, a convergence of events led me to wake up once and for all: the solo trip out of town and preemptive power safety shutoffs that I spoke about in the Introduction, two nearby wildfires, and an important speech. One night, I woke in the night to the first scent of annual wildfire smoke, feeling scared, trepidatious, and very much alone (despite my husband sleeping soundly next to me). When my husband went out of town for a few nights, I paced in my bedroom after the kids were asleep, wondering what on Earth I would do if there was a fire and we had to evacuate. In the daytime, I'd find myself standing on our front walkway, my ears perked, my eyes fixated on tree branches whipping around in hot, windy weather, my skin sensing danger in the air.

But the pivotal moment that sparked my true e-piphany came in September of 2019. Greta Thunberg was speaking at a UN climate change conference, and I was listening on the radio. I'd just dropped off my kids at their respective schools when I heard her sharp words pierce through my morning brain fog: "You've stolen my dreams and my childhood with your empty words! I should be back in school on the other side of the ocean. Yet you all come to us young people for hope. How dare you!"

I felt something delicate in my heart region break like porcelain. Strong, unapologetic emotions burst past the dams and constructs I'd built up for so many years, and I cried and cried. There was no stopping this flood, even if I had wanted to. (And, as we all now know, it's important to allow emotions to run their course, and pay attention to them.)

Tears wet my cheeks on my drive home, continuing as I pulled up in front of my house. Once inside, I tumbled in a heap onto the couch and a new wave of tears cropped up, a muddled explosion of grief, guilt, shame, anger, and sadness. Why was it that I needed a young teenager to talk sense into me?

As I heard Greta's angry words spew across the airwaves, I experienced a two-fold response. On the one hand, I identified as the parent receiving harsh words from their child, feeling guilty and complicit in the ways we as adults had failed her generation. But on the other hand, I identified with young Greta. In my bones, I felt a resurgence of youthful anger and betrayal because *I* had felt as she did as a teenager. I, too, had paid attention and taken action, often questioning why few others seemed to care. Greta's words struck a double chord for me.

Since then, I've done some of the deeper reflecting work that you will do in this chapter and the next. I now know that Greta triggered a latent but still very much alive part of myself—my caring, devoted youth activist self, a part of me that cared deeply for all living creatures, that rescued injured bunny rabbits on the side of the road, that whispered to myself "rest in peace" whenever I drove by roadkill, the one who plastered "against animal testing" signs in the makeup section at my local pharmacy, and published an op-ed in the local paper about the cruelty of the fur industry.

Greta's words forced me to look squarely in the face at my suppressed values, and therefore my complicity in this crisis. I hated to admit it, but my activism began atrophying in my late twenties. Unwittingly, I'd become like those same adults who I'd once judged as I chose instead to focus on other matters in adulthood. Matters that seemed trivial in light of my e-piphany, yet were understandable: grad school, career, finding a life partner. During that time, my ecological values had been relegated lower on my list of values. And this stayed put as my adult stressors mounted in my early thirties: marriage, mortgages, babies! My e-piphany was coming to grips with my years of inaction, and the realization that I could choose differently at this urgent juncture.

That e-piphany (thank you, Greta!) has proved to be a defining moment in my life. Since that day, I haven't been able to look away from the climate emergency. As I shared in the Introduction, I'm engaged on all fronts now:

professional, personal, political, spiritual, and creative. I continue to act and to pay attention in all these realms because it's what feels natural and vital. I'm continually identity sculpting and contorting into new vulnerable shapes.

Since that pivotal moment, I've shifted my private practice to include ecotherapy and climate aware therapy. I've said "yes" to climate-related presentations and podcasts, "yes" to teaching about climate, and "yes" to writing a book. Yes, yes, yes! Since my e-piphany, there's never been a question whether or not to stay engaged; I now feel compelled to do so. This compulsion is continually reinforced as I raise my kids, even as I've had to learn how to set healthy limits for myself. (And, yes, even as I've bulldozed my limits and regrouped.)

In my personal life, my husband and I have been able to implement some of the greener lifestyle changes that feel hugely relieving on a day-to-day basis: buying an electric car and an electric bicycle, installing solar panels to electrify our home, using fewer single-use plastics and recycling what we do use.

Keep in mind that e-piphanies and aha! moments are not the only avenues to pivotal action. There are so many other rich access points, such as dreams, intense life experiences, vision or other spiritual quests, psychedelics, living abroad, volunteering, motivating conversations, among others. Identity sculpting has many access points and no right way to do it, which makes it very accessible.

Since the focus of this book is on engaging the Parent Club, I've focused on adult e-piphanies and aha! moments here. But, of course, kids experience these all of the time too and can also share their stories. Perhaps they might even inspire you to act or join forces.

As parents, our actions—or inactions—are a significant form of modeling. How we respond to this moment matters. Do we roll over and play dead or ignorant, or do we rise up, together, to demand our governments and world leaders do something?

The exercise below is a fun way to integrate some of the content from this chapter. It can be done with kids too, if it's simplified and/or translated into kid-friendly terms.

CREATIVE EXERCISE

VISUALIZING THE FUTURE WITH A SOLUTION-FOCUSED MANDALA

Materials:
- Paper, journal, notebook, art journal, construction paper
- Colored pencils, markers, gel pens (feel free to use wet media if you'd like)

Setup:
Trace a mandala onto paper by drawing around a circular object, such as a plate, bowl, or lid.

Prompt:
Inside of the circle, draw a picture of the issue/problem/concern that you are focused on.

When you're finished, draw an outer ring with plenty of space between the two circles. Choose three or more ways that you plan to contribute to or tackle this issue no matter how big or small. Divide the outer ring into three or more parts accordingly and write or draw those specific contributions in each segment.

Finally, outside of the mandala, draw a picture of what the world might look like if you woke up tomorrow morning and the pressing issue that you were concerned about was resolved(!).

Process:
As you draw and write, notice without judgment any feelings, thoughts, and sensations arising in your body.

Product:
Spend some time reflecting on this exercise and all of the tools that it offers you in forging ahead and constructing your parent identity. Discuss or journal about it. Consider sharing it with someone else.

Reflection questions:
- How does it feel to look at your artwork?
- Is there movement/flow/a sense of radiation or other directionality?
- Is there a message in the art?

PART 3

Community

NINE

From Alpha to Boomer to Z

Strengthening Intergenerational Work as a Way Forward

"We come from different places, but we're fighting for the same cause."

–High school student

Have you ever been to a silent dance party in which everyone has a pair of headphones on and is listening to one of several color-coded stations? You can look around and see what others are listening to by the color that pops up on their headphones. When you see someone dialed into a different color than you, and watch them dancing to a completely different song, it can feel humorously disjointed.

All too often these days, it can feel as though we are siloed like this, particularly in the climate space. Everyone is listening to their own music, establishing their own rhythms and moves, and part of a different experience. It's fine for a night out, but it's not healthy or productive for a climate emergency.

Of all the divides I see in the climate space, the intergenerational one feels like the most overlooked yet ripest for action. I fruitlessly tried to locate a local intergenerational climate group, but it was surprisingly difficult for me in the climate-forward Bay Area. The best ones I could find were more general climate groups, or a network of parents concerned about various social issues, such as police brutality, ICE detention centers, and climate change. The Seattle-based group Climate Action Families is on the mark and seeks to "empower an intergenerational climate justice movement of youth, families, and friends building collective power to protect the people and planet."[1]

To be sure, there are national and international groups with this focus, such as Our Kids' Climate and Parents For Future, with grassroots groups in at least twenty-three countries. But we need more intergenerational events, more Parent Club participants, more traction.

To that end, this chapter will take a deep dive into creating an intergenerational climate movement with the thinking that this roundup can apply to other social movements as well. As you'll see, there are plenty of opportunities and inroads for how to collaborate more effectively across the age spectrum. But in order to address the problem, we must first take an honest look at our intergenerational wounds and dynamics.

Understanding Generational Divides–and How to Bridge Them for Stronger Social Movements

The picture in my mind that I keep coming back to is a person whose arms and legs are moving out of syncopation, separately from one another, and cannot function efficiently. It's frustrating, exhausting, tedious, and ineffective.

As technologies inevitably develop from generation to generation, these stamp impressionable young minds and shape their development, creating natural rifts between generations. Life is vastly different on either side of the printing press—or the smartphone! Lifestyle changes inevitably result from human progress and innovation.

In some Western democracies there's a dogged belief that economic and technological progress is good and that life will improve for future generations. But what about when the trajectory is not looking so good, one that is looking bleaker, less stable, inequitable, and more unpredictable?

There is far less certainty in a world riddled by intensifying flooding, drought, wildfire, storms, and other natural disasters. Compounding our climate emergency are anti-environmental economic interests such as oil and gas, industrial manufacturing, aviation, and factory farms. And the continued lack of government regulation to protect our environment and public health is a slap in the face (or the formaldehyde cherry on top, for that matter). As parents, grandparents, and adults, the least we can do is take some time to really see it through the kids' eyes.

"This isn't fair!" "Why don't you act?" "Why do you think you know best—when clearly, you don't?" These are some of the voices of our children. They are pissed, saddened, and frightened. And of course they are. Our kids don't want to grow up in more pollution, with less biodiversity, less clarity about if they want to have kids of their own someday, and moral hangups about air travel.

An eighteen-year-old Bay Area high school student told me that growing up as a young person right now is vastly different from what it was like for previous generations. "We can't really plan for our future," she said, "because we don't know what the world's going to look like in fifty years. Older adults . . . had more stability and could just be kids. There's a visceral feeling of urgency, fear, and uncertainty now. Older adults don't often understand the mental load that this has on younger people."

It's hard to know what to say and not to be defensive. Swedish youth activist Greta Thunberg has frequently expressed her frustration by referring to adults' all talk and no action habit as "yadda yadda yadda." Note for older adults: Gen Zers want action and results, not touchy-feely compliments about how they will be the generation that will solve climate change! These differences have paved the way for siloing among climate organizations. As a result, there are youth-led organizations, elder-led organizations, and parent-led organizations, and many other ones catering to specific age demographics.

Young people today are experiencing intergenerational injustices: unfair existential burdens and "intergenerational theft" placed on them due to the consequential actions or inactions of elder and past generations. These injustices are further exacerbated by their exclusion from decision-making spaces, or voting, a lack of access to resources and green space, and profound environmental racism.

Some adults find the comments made by younger generations to be spiteful, ungrateful, and hurtful. Some are afraid of their wrath. A subset of Boomers feels real guilt and shame about the state of the world they are leaving behind for their grandchildren and great grandchildren.

The Boomers I typically encounter through my climate work fall into this camp. Those who have the courage to stare down their fears and regrets have a deep determination and commitment to use their talents and tools to fight for the climate cause while they still can—just as passionately and urgently as youth activists. While their modalities look different—more Zooms and listservs, fewer TikToks and Snapchats—these Boomers share dedication to the same cause. This is exactly where the rubber can meet the road and we can start to build traction through apology, forgiveness, and compassionate collaboration.

While everyone is entitled to their feelings, stewing in these tensions is a grand waste of time. After all, if we remain divided, it undermines our ability to act and benefits oil and gas companies and those with power who are profiting from ignoring climate change. We need to shift our blaming habits and emotional reactivity toward more constructive and empathetic responses. Let's begin this process with a solid cognitive reframe—examining the variety of strengths and skillsets held by different generations of human beings. Acknowledging our strengths and abilities can fuel so many social movements!

STRENGTHS AND SKILLSETS OF EACH GENERATION

What does each generation bring to the table?

Kids:
- Fresh perspectives and ideas
- Optimism and hope
- Open-mindedness and enthusiasm
- Physical energy
- Time

Teens:
- Tell it like it is; real talk
- Notice and call out any hypocrisy
- Conviction to act
- Willing to take risks, and act courageously
- Peer influencers
- Technological and social media savvy
- Culture shapers
- Message amplifiers

Parent Club:
- Strong multitasking abilities (well, at least some of us)
- Badass vulnerability
- Life experience

- Mentorship
- Opportunities for intergenerational linkage

Elders:
- Abundance of life experience, knowledge, education
- Long-term perspective
- Circle of influence; connections
- Mentorship
- Story-holders: experience, history, adaptation, and culture
- Time
- Resources
- Civic engagement—most engaged group; highest voter turnout
- Motivated by legacy, contribution, and death

Look at this rich bounty on this table! Clearly, intergenerational collaboration fosters connection, resilience, solidarity, and commitment, qualities that will buoy us during these uncertain times. Through this lens, intergenerational collaboration is a no-brainer.

Remember: Age Is Relative!

As we approach these challenges, it's important to remember that age is a marker, but it does not define you. Age can also be a mindset, and is in a sense more fluid than we realize. In Aboriginal culture, every three generations there's a reset in which your grandparents' parents are classified as your children. At any time, you hold multiple relational contexts. Kinship moves in cycles that renew and bind relations, which is helpful for perspective-shifting and intergenerational repair. When we understand aging along a spiraling spectrum rather than as prescriptive categories that mean x-y-z, then we can start to see bonds—instead of lonely chasms—between us.

While reading Dr. Seuss's classic children's book *The Lorax* to my youngest daughter one night, a startling truth hit me. At different phases of our lives, we might resemble any of the three characters in the book: the child, Once-ler, and Lorax. It dawned on me that Karpman's drama triangle shows this connection, through the roles of victim, persecutor, and rescuer. In other words, we inhabit each of these roles at different times in our lives. For example, as a child, or rescuer, you might have been curious and receptive to storytelling and learning, and approached life with hope, empathy, and a desire to do good. But at another stage of life, you might have been like the Once-ler, the persecutor, your greed having overrun your values, leading you to cause harm to the environment or stay in MA rather than do something about it. Perhaps now you are like the Lorax, who stands up and advocates for a cause (in this case the Brown Bar-baloots and Truffula trees). If we can remember that these roles are fluid, this can help us to resist all-good or all-bad constructs, see past ageist beliefs and othering, and bind us together in our humanity. Our identities, after all, evolve over time.

So, What Does Intergenerational Work Look Like, Anyway?

What if we learn to pause, to listen, and to respond rather than lash out? What if we use our emotions to empathize and connect? What if by repairing relationships we build a stronger foundation for enacting change? Generation-specific organizations exist because groups often form around commonalities, and members can feel energized and validated by shared perspectives. These intergenerational organizations take a different approach, tapping into our diversity as potential:

- Generations United (www.gu.org/) is a national organization dedicated to improving the lives of children, adults, and older adults through intergenerational collaboration that have inspired projects all around the US.
- HelpAge International (www.helpage.org/) has piloted the Uniting Generations for Climate Action project in Nepal and Uganda.

- Research Institute for Future Design (www.souken.kochi-tech.ac.jp/seido/index.php) is a participatory decision-making process started in Japan that considers how our decisions today will affect future unborn citizens (it was inspired by the Haudenosaunee Seventh Generation Principle; that is, to consider the well-being of the next seven generations in decision-making).
- Climate Mental Health Network (www.climatementalhealth.net/) is an intergenerationally focused nonprofit that includes a Gen Z advisory board.

These are but a few examples of budding intergenerational work. Such initiatives build mutual respect and a sense of collective responsibility, strengthening intergenerational solidarity. I expect that momentum in this sector will grow in the years ahead. But in the meantime, perhaps, we can learn about how to better collaborate as independent organizations in the broader climate ecosystem.

Models for Moving Forward

Remember that although many US social movements—the civil rights movement of the 1960s, Vietnam-era anti-war protests, the LGBTQ+ and Black Lives Matter movements—have been fueled by strong energetic young people demanding action, they were also sustained by elders who contributed their life experience, knowledge, and tenacity.

The story of Standing Rock, North Dakota, had multiple groups—elder tribal leaders and their allies—coming together to protest the Dakota oil pipeline. In doing so, they achieved a temporary blockade. This is also the story of the US Civil Rights Movement, in which established organizations like the NAACP merged with younger, radical organizations like the Student Nonviolent Coordinating Committee, both challenging each other to a stronger end. Elders like labor unionist A. Philip Randolph came in at critical moments to lend advice and support. While there was tension between the groups, there was also connectivity, and this energy fueled successful public resistance like the March on Washington and the Montgomery Bus Boycott.

Just imagine all of the many different climate organizations in the world finding a sense of connectivity in the broader climate movement and what that could mean for synchronicity, productivity, and impact. Like an orchestra comprised of different instruments, each playing its own tune and coming together as one to produce a beautiful melody. Each musician is aware of their part in the musical score and sticking to it in service of collaboration. At times, one instrument may carry a solo tune, but then it switches, and resettles back into the depths of orchestral arrangement. That's the potential of a cross-generational collaboration.

As you will soon see, if we stand together linking arms, a lot more can happen—and faster.

School-Based Intergenerational Climate Collaboration

There's an urgent need to develop school curriculum that addresses climate change. For decades we've left teachers scrambling to improvise and respond to kids' questions as they see fit. This vacuum has led to inevitable political hijacking, with states such as Texas weakening climate science education guidelines and risking future generations being unprepared for the world they will inherit.

Developing school climate curricula is a concrete task that naturally engages multigenerational perspectives and group collaboration: teachers, administrators, parents, and students. From an ethical standpoint, these curricula ought to ground teachers in the basic science of climate change, addressing global and local impacts (including health), exploring climate solutions, environmental justice, and equipping kids with skills in emotional resilience.

Let's consider my conversation with Bay Area youth climate activist Ella Suring, who was part of a four-woman team (including a retired teacher, a parent, and a school board member) that helped pass and receive funding for the Berkeley High climate literacy resolution in 2021. In middle school, Ella had often felt anxious and alone because nobody was talking about climate change, even when it manifested in annual wildfires that raged in surrounding communities. She told me that many of her peers in middle school and later in high school were busy, distracted, or just didn't prioritize climate in the same way that she did. Her sense of anxious aloneness culminated in 2020 when she was fifteen years old, during an intense wildfire that overlapped with the pandemic.

Attending meetings, collaboratively planning a curriculum, and seeing how adults consistently dedicated their time and skills offered her hope during a dark time. Although there were many hoops to jump through to pass the curriculum, she had the support of an intergenerational team, and it helped that one member had experience passing similar resolutions locally.

Ella said that she doesn't harbor anger or resentment toward older adults and doesn't think that mindset is helpful. "I can learn a lot from older adults," she said. "They are great at communicating, they have experiences I don't, and their developmental and educational viewpoints in this case were essential."

On the opposite coast, in Pennington, New Jersey, art teacher Carolyn McGrath cochaired the student-led Youth Environmental Society (YES). YES was responsible for researching and presenting a climate action proposal to the Board of Education in 2022. As part of the proposal, a district-wide climate action committee was formed by the Board of Education that included students, teachers, administrators, Board members, and parents. This committee used a student-created climate action proposal as the roadmap to a developing a comprehensive district climate action plan. Training teachers to teach about climate change is one of their priorities. YES has also been involved with Schools for Climate Action in Washington, DC, advocating for national climate education and youth mental health support.

During my interview with McGrath, she noted that while students feel the urgency of the issues, most schoolteachers, administrators, and parents do not. She described a certain type of "magic realism" that pervades the adult sphere. She said that hardly any adults in their educated, upper-middle-class school district openly talk about climate change, but she did note that a handful of engaged parents participate on the climate action committee. If more parents from the Parent Club could engage like they do, things would move faster. That's why it's our job to pitch in.

Getting Legal with It

Children and adults are also busy collaborating in international courts. Julia Olson, a Portland-based lawyer, launched the nonprofit organization Our Children's Trust in 2010, bringing young people front and center as plaintiffs in the groundbreaking climate case, *Juliana vs. U.S. Government*. The case was filed in the US District Court for the District of Oregon in 2015 on behalf of twenty-one kids, some as young as eight years old. The kids' interests and futures were represented by lawyers, psychiatrists, and other adult professionals. This case has been subject to political ping-pong, repeatedly squashed by the US Department of Justice. In 2025, the US Supreme Court declined to hear the case, effectively closing the case. However, this innovative legal battle has reshaped the climate fight and paved the way for similar litigation.

In 2023, *Held vs. Montana* made international headlines when sixteen Montana youth sued their state government in an effort to protect their constitutional right to a healthy environment, life, dignity, and freedom. With the help of Our Children's Trust, they argued that their health and futures are jeopardized by government decision-makers who supported continued fossil fuel extraction. This intergenerational team of youth plaintiffs and their adult attorneys won what is considered to be the first constitutional climate trial in the US, the case setting a critical precedent. The ruling affirmed the scientific reality of human-caused climate change, specifically in how it affects the lives of youth plaintiffs in the state of Montana.

Another successful climate lawsuit was filed by young people in Hawaii against the state's Department of Transportation. In the settlement, the defendants have agreed to achieve zero emissions by 2045 and establish a volunteer youth council to review and provide feedback on their plans.

As a result of these landmark cases—all examples of intergenerational collaboration—a message slowly gaining traction around the globe is that youth voices matter and can make a tangible difference in political outcomes. Youth-adult coalitions have the potential to enact positive changes for human and planetary health and well-being.

Rev Up Your Parent-Led Advocacy Engines

First things first: parent climate communities kick butt! I'm heartened to know of a number of parent-centered organizations that provide support and solidarity as we all try to stay engaged. A few examples include Moms Clean Air Force, Climate Mamas and Papas, and Science Moms.

Doing something on the climate front as part of a community feels relieving because there is suddenly a channel for your emotions to move through. When there are other people around you don't feel as paralyzed and helpless. But more than that, when you feel you are part of something bigger than yourself with a shared vision it's straight up empowering.

Groups also create accountability and a container for your climate work. They provide places to return to and see progress and growth—or nurse setbacks and grieve together. As we discussed in chapter 8, developing your voice and identity and translating this into activism and advocacy work is a fulfilling and relieving mission. Joining a group will propel you in that direction. You need not have experience to apply!

A board member of Chesapeake Climate Action Network told me that there's a formula to how their group collaborates: adults provide social connections, funding, and ideas, and the kids "run with it" and manage a lot of the day-to-day operations. She noted that kids have drive and physical energy to an extent that most grown-ups do not. "I've done my arrests, so at this stage in my life I meet with people and use my power and influence," she told me. There's a sensible mind-body synchrony here, in how adults and youth contribute what resources, skills, and experience they have in a complementary way.

Taking action can actually be a sweet spot of connection for parents and kids, stimulate more conversations, and build hope and trust in the process. The idea is to find support and collaborate efforts with others doing the work. It could also be offering to be of service to youth climate groups, including fundraising, fostering important connections, or serving on advisory boards.

> **ENGAGING AS PARENTS AND CHILDREN**
>
> **Luz E. "Lucy" Molina, a longtime resident of Commerce City, Colorado, has endured years of poor air quality and contaminated water and soil. Lucy lives in what she calls the "Bermuda Triangle" of polluters: close to the Purina factory, Suncor, and Xcel Comanche natural gas plant. In an interview on the *Hot Mamas* podcast (an excellent catalog of mothers involved in the climate sector), Lucy calls her hometown "environmental vomit"[2]; it's a sacrifice zone. Bloody noses, asthma, cancer, diabetes, and migraines are common in her community and have affected her own family. As a single mom of two, her health concerns prompted her to step into action. Lucy started knocking on doors, speaking up at city council meetings, and testifying at commissions. Her work as a community organizer has helped advance a settlement with Suncor to pay for air quality violations. She's part of the "EcoMadre" movement at the nonprofit group Moms Clean Air Force, which brings Latina and Indigenous mothers together. As she's shared: "This is my home. As a mom, it's why I stand up. We have a responsibility to protect our planet – it needs to begin in communities**

> like ours that have been impacted with environmental racism and injustices."³
>
> How parents engage on social issues makes a lasting impact on children's values.
>
> At the ripe age of eleven, Lucy Molina's nephew, Elijah Molina-Moton, was lauded as Climate Hero of the Month from 350 Colorado, a grassroots movement for 100 percent renewable energy. He began speaking publicly about his experience growing up near Suncor in a sacrifice zone at workshops and conferences. He shared his story about managing his asthma and his hopes for a world where the air we breathe is not full of pollutants. Elijah grew up watching his aunt speak out and take action to make a difference for their local community on this issue, and she undoubtedly inspired the next generation.

Rev Up Your Parent-Led Activism Engines

At protests ranging from women's marches to civil rights, from March for Our Lives to Black Lives Matter, from anti-war demonstrations to climate week, photographs in the media often capture multigenerational participation: babies in strollers, kids sitting on shoulders, and teens and adults of all ages marching and making noise. One way I differentiate advocacy from activism is thinking of advocacy as speaking out with your words and activism as speaking out with your body (although they overlap a lot).

For folks who were youth activists, activism means returning to your roots. But for others it might be something completely new. There are many family-friendly rallies and protests that can feel inspiring to be a part of or support from the sidelines. Of course, parents have to weigh safety before putting their bodies on police lines and subject to arrests, tear gas, and the uncertainty of protests. Whatever form of activism you take sends a powerful signal to those in charge that can't be easily ignored.

Let's take closer look at intergenerational activism at Standing Rock. For the climate movement, this was a pivotal moment—a national movement moment—as tribal activists demonstrated to the rest of the world what the power of this activism can look like. In 2016, a grassroots opposition movement to the Dakota Access oil pipeline construction project in North Dakota captured the attention of the nation and threw a wrench into oil and gas development. An elder member of the Standing Rock Sioux Tribe, LaDonna Brave Bull Allard, along with her grandchildren, set up a water protectors camp in effort to protect a sacred site, the Missouri River. Young people from the Standing Rock Sioux reservation as well as from surrounding reservations organized protests at the pipeline site to prevent it from advancing further, locking themselves to machinery as a form of civil disobedience. Thousands of activists poured in, including from hundreds of US tribes and from around the world, joining them on the frontlines. The pipeline project was delayed, denied, and temporarily shut down. Although eventually completed in 2017, the strength of this protest, which gained significant media attention, is proof of intergenerational might. Activism that brings people together has unfettered energy and potential; if each of us rises, business as usual cannot go on.

Another inspiring example of intergenerational action was a group of mothers who called themselves Las Madres del La Plaza de Mayo that formed in the late 1970s to protest the disappearance of their adult children under Argentina's military dictatorship. Despite the risk of public demonstrations, mothers protested for years outside the presidential palace to draw attention to human rights violations and to demand the return of their children. Their protests helped with the later sentencing of military commanders for their roles in the deaths and persecutions of innocent people. At critical times, parents can link arms and show up in an effort to protect children, and point fingers at injustice and oppression. We can do this too in the context of today's stressors.

Another intergenerational story that highlights the power of the youth climate movement is the Green New Deal. In 2019, congressional legislation emerged focusing on advancing clean energy initiatives with respect to environmental justice. How did that come to happen? The Sunrise Movement, a robust youth-led organization headquartered in Washington, DC, staged a sit-in the year before to advocate for a Green New Deal at House of Representatives Speaker Nancy Pelosi's office. This proved a climactic moment not just because of the courageous activism of a generation, but because of a successful generational baton-pass to rising political star Alexandria Ocasio-Cortez, who joined their sit-in and carried their cause to Congress. Ocasio-Cortez championed the Green New Deal, using her newfound political power to advance it. Later that year, she and Rep. Ed Markey from Massachusetts cosponsored the bill in Congress, although it eventually stalled in the Senate. Sunrise also helped to make the Green New Deal a central issue in the presidential campaign in 2019. The Green New Deal offered a strong vision for a new, more equitable society, eventually spawning into a diluted but effectively passed climate bill called the Inflation Reduction Act (2022). Even if this act did not have everything that environmental advocates were hoping for—and despite its renunciation under President Trump—it was nonetheless the single largest climate legislation passed in US history. This is how different generations intersect for good to move climate legislation forward.

CLIMATE ACTIVISM

When embedded into family cultures at an early age, climate activism is picked up by little kids' antennas. Prominent Indigenous youth climate leaders Xiye Bastida and Helena Gualinga came from families of climate activists. Bastida was born and raised in Mexico in the Otomi-Toltec Indigenous nation, and her parents met at the UN's First Earth Summit in 1992. What's she up to now as a young adult? Well, she's an organizer for Fridays for Future and the cofounder of

> Re-Earth Initiative, an international youth-led organization that focuses on highlighting the intersectionality of the climate crisis. Meanwhile, Gualinga, who grew up on the Ecuadorian Amazon frontline when her Sarayaku Indigenous community was fighting an oil company for trespassing on their lands, is also making big moves. She has become a community spokesperson, speaking out against her government, attending protests, and cofounding the organization Polluters Out.

Expand Your Creativity in a Generational Context

There are countless ways to engage both creatively and intergenerationally. Here are a few points of inspiration:

- Oakland-based mom Lil Milagro founded a Dungeons & Dragons group at the elementary school where she worked. This group, eventually spawned a youth-led climate nonprofit, the Mycelium Youth Network. She helped create Gaming for Justice, an interactive video game set in Oakland that gives kids a chance to solve real-world problems such as pollution, deforestation, and police brutality. She noted that such video games can be educational, empowering, and build social connection.
- Becky Mailer, Maya Ward, Chryso Chellun, and Emma Powell cofounded Mothers Rise Up in the UK, which organizes multimodal creative protests that draw hundreds of participants of all ages, including children, incorporating singing, dancing, music, theatrics, and visual arts. Think: climate flash mob. Brilliant. A main focus has been protesting outside Lloyd's of London and demanding that it stop bankrolling the fossil fuel industry.
- Climate Change Theatre Action is an American-Canadian collaboration that produces short plays about the climate crisis timed to coincide with annual COP meetings. The group hosts workshops for

young people teaching them how to create climate change theater of their own. The group is housed under the broader Arts & Climate Initiative (artsandclimate.org).

- The Climate Museum sprung up in New York City in 2018, offering rotating exhibits at different museums and spaces to educate and engage folks of all generations. There are pop-up community events, internships, climate action leadership programs for high school students, and even opportunities to write to Congress. In 2029, the Climate Museum will open its permanent home.

Now that you're feeling inspired, get ready to harness your creativity for good use! What follows are some fun ways to dip your family's toes into climate-focused creativity that will open up emotions, questions, attitudes, and build resilience. Since the name of the game in this chapter is intergenerational work, I encourage you to try this out.

CLIMATE-FOCUSED CREATIVITY

Let's get crackin' with climate-focused creativity!
- *Collage poetry.* Assemble images and words that illustrate how you are emotionally impacted by climate/environmental crises. You can be specific to where you live: How do you feel about wildfires? Unprecedented heat waves? Living near a toxic waste site?
- *Puppet play.* Create a representation of a "hope" puppet and a "doomsday" puppet using brown paper lunch bags, paper shapes glued to popsicle sticks, or using fabric and hot glue or sewing. Next, put on a puppet show or hold a dialogue between these two characters. What story played out? What ideas come forth? How might a different scenario play out?
- *Emotional anchor.* Create an anchor that will see you through hard times and maintain resilience in all types of weather. Try using 3D materials like clay or sculpture, cardboard, popsicle sticks, branches, etc.

- *Painting flowerpots and planting seeds.* A family project for learning about care, empathy, patience. (Have you ever read the classic book *The Carrot Seed?*)
- *Intergenerational commitment tree.* Locate a tree in the community for this project. Cut out same-size paper leaves and place these with a few pens on a small table or nearby tree stump with a bowl of 8" pieces of string or twine. Each leaf represents an ecological commitment that a community member is willing to make. You write the commitment on one side of the leaf, and perhaps your name and age on the other, or it can be anonymous. Create a sign for the project as well as offer instructions and an explanation of what this is all about. Be sure to visit regularly and read the leaves. You could also host a community gathering to share this experience.

Having Those Difficult Conversations . . . Respectfully

As we covered in chapter 7, there are skillful ways to communicate so that people actually listen to one another and engage in meaningful dialogue. Climate scientist Katherine Hayhoe underscores the importance of everyday conversations in her 2018 Ted Talk: "The most important thing you can do to fight climate change: talk about it."[4] More recently, Hayhoe has pointed out how climate change has become a polarizing identity issue, and says that just saying the words "climate change" can put people on the defensive. Intergenerational climate conversations can be particularly challenging and daunting. But we must find ways to connect and converse despite our different points of view.

Without further ado, try out the exercise in the box that follows. You can do this as a parent and child with an elder adult, or you can support your child in doing this, or any variation. Keep in mind that you'll need to shift the language around to make it more kid friendly. Isn't the lesson we so often tell kids is that the only way to get better at something is to practice and not to give up?

SPARKING AN INTERGENERATIONAL CLIMATE CONVERSATION

To start your intergenerational climate conversation:
- Grab a notebook and pen, or set up an audio recorder.
- Review your NVC (as mentioned in chapter 7).
- Choose somebody of a different generation to interview or spark a climate conversation with (ideally someone that you have a connection to in some way).
- Ask permission to chat: "Do you have 15-20 minutes to talk about something that's very important to me?"
- Explain your mission: "I'm interested in hearing your perspective on the changes that are happening around the world, like record-breaking heat waves, more frequent and intense hurricanes, loss of species, that sort of thing."
- Invite an exchange: "I'd love to share my perspective with you too if that's something you'd be interested in. I know how we can have a respectful discussion that could benefit both of us and help us to understand each other better."
- Collaboratively set the ground rules (e.g., length of time, appropriate questions).

Sample interview questions:
- Have you noticed any changes to your local weather, lands, water, air, and animal and insect species? If so, how do you feel about that?
- Has your attitude changed in your lifetime?
- Did you ever have an e-piphany?
- Do you want to learn more or do more about this issue?
- Are there other folks in your life who show an interest in this topic?
- What's your biggest concern about the future?
- What do you feel most hopeful about?

Journal or debrief with the interviewer and later with a climate buddy:
- How was this experience for you?
- Did you learn anything new?
- Did you have anything in common?
- What points did you disagree on?
- Were you both able to participate respectfully?
- Would it be possible to collaborate or just stay in touch in a way that feels promising?
- What did you find challenging about this exercise?
- Is there anything you might do differently next time? Or work on before your next interview?

Congrats on making it this far and having the courage to talk climate! A big standing ovation for you! I'm going to briefly acknowledge here a few other promising avenues for fostering healthy conversations.

- **Climate cafés** are informal spaces where people can come together, and share and listen to each other's feelings and experiences. Hosting or attending intergenerational climate cafés can be an exceptionally rich experience. But for a creative twist, consider holding a family- or community-based climate café. This could be a family, multi-family, or intergenerational community event. It could be recurring annually, semi-annually, or quarterly. Of course, it will be important to set simple group rules in order to build a safe and nonjudgmental space. (See my website for guidelines.)
- **A good creative brainstorm.** The possibilities of connection and expression here are infinite with this. Here are some ideas:

    Writing music, or playing musical instruments (e.g., drumming, band, improv)
    Singing
    Curating a playlist

Painting a mural

Other forms of art-making (e.g., natured-based materials, mixed media)

Reading poetry

Spoken word

Sharing photos, videos, or digital artwork

Storytelling

Puppetry

Sharing journal entries

Performing dance

Performing a play, role-playing, or playback theater

- **Family-Centered Ecological Values.** The most important thing is for folks to be able to express themselves in a way that feels comfortable or enticing to *them*. Being seen and listened to by others can feel hugely relieving, decrease a sense of loneliness or despair, and promote hope and a sense of community.

- Other ways to connect with family around ecological values include learning together/education, planting a garden, and creating a family plan for civic action (see chapter 3). Appendix D includes a list of community initiatives that support access to nature for low-income and BIPOC communities.

- **The Friendship Bench** is a promising model for facilitating cross-generational conversations designed by psychiatrist Dixon Chibanda at University of Zimbabawe. Zimbabwe has a shortage of mental health professionals, and The Friendship Bench recruits and trains elders (most often women) to provide informal counseling to fill in those gaps. Studies have shown remarkable improvement and symptom reduction for participants, even when compared with pharmaceutical treatments. This model is beautiful because not only do elders offer much-needed emotional support to struggling parents or young people, but the elders are enriched through the experience of social connection and feeling valued. Since the pandemic, Friendship Bench DC is building on this model to uplift marginalized people of color in the local community.

The most important thing is for folks to be able to express themselves in a way that feels comfortable or enticing to them. Being seen and listened to by others can feel hugely relieving, decrease a sense of loneliness or despair, and promote hope and a sense of community.

So, How in the Heck Do We Build an Intergenerational Climate Movement, Anyway?

We've established that intergenerational teamwork is the way forward. We've also looked at some successful examples of organizations and movements. But we've got to forge our own paths in our homes and communities at the end of the day. To contribute to a powerful intergenerational wave of change, there are many steps we can take as individual Parent Club members.

Be a Climate Connector. Community offers comfort and connection. Bringing people together rather than alienating them is important—and that includes youth and elders. In 2023, I was invited to attend a local retreat with a nonprofit called NorCal Climate Elders. Throughout the day I fielded questions such as "You're not one of us—what are you doing here?" I told them that I was interested in supporting intergenerational climate work. Although the group was humorously self-disparaging ("Oh, we're just a bunch of old folks"), it was clear how motivated and determined they were. There were several deep-dive experientials so that participants could process climate emotions, support one another, and even think from other generational perspectives. There were lectures, team building whiteboard exercises, and, as a parting gift, each participant received a redwood tree to plant.

Throughout the day, enthusiasm for my participation mounted, and the next morning I had a number of participants saying, "We want to hear what you have to say . . . want to take the mic?" As I was bidding adieu, one member asked me to propose an idea for an intergenerational retreat next year. He seemed to read my mind (or facial expression), which expressed the ambivalence of both wanting to do something but feeling too overwhelmed. "All you have to do is come up with the idea," he told me. "We can do the rest. We have plenty of time and resources to make it happen." This illustrates the potential of cross-generational collaboration to the tee. (Incidentally, the following year, the same group sponsored a Future Design Retreat shifting more definitively toward intergenerational work.)

On the other side of the age spectrum, I spoke with Climate Mental Health Network's Gen Z adviser Maksim Batuyev. I was struck by his openness and eagerness to collaborate. During our conversation, he noted that the climate mental health nonprofit I'm involved with would benefit from greater youth involvement. I concurred. He said: "Y'all have all of the expertise in the world, but you must be tired from the work that you're doing. You could use young people who bring fresh ideas and energy." For a second, I think that he worried he'd offended me, but I quickly replied, "You're absolutely right. The truth is that I'm a tired parent."

Batuyev is the director of Climate Café LA, which brings together young people from across the globe into climate café forums. He has a generational reach that we don't. Considering that climate psychology has a "marketing problem" (as he puts it), this reach is key. Maksim shared that many Gen Zers are distracted and not engaged in climate, and certainly don't identify as climate activists. He believes there's a huge potential to reach this swath of the population. We've discussed collaborating on future intergenerational events, such as a climate poetry workshop.

Follow your child's lead. If your child is already engaged in a cause, you could ask for permission to attend a meeting with them, offer support, or offer to volunteer with them. Be a vocal, proactive ally.

Greta Thunberg's father, Svante, initially wary of media attention and its implications for his daughter, found a path forward in supporting her activism. When she started skipping school on Fridays to protest outside the Swedish parliament, he was close by. Her actions seeded the School Strike for Climate movement. By supporting her, of course, Svante indirectly helped awaken the international community to the cause of climate change and need for a rapid response.

Wield your existing profession or platform. You carved out your climate identity in chapter 8, but you can certainly pair this with intergenerational work such as mentoring, cocreating projects, or consulting with youth. The Environmental Defense Fund hosts the podcast *Degrees*, which is specifically focused on how folks grow their careers in climate.

Lil Milagro, who I mentioned earlier in the chapter, heard kids' worries about wildfires at the after-school D&D club and asked them if they'd like to form an organization. The answer was yes, and the kids began piloting climate-learning programs based on their questions and experiences. Lil's approach has been collaborative and exploratory rather than top-down. She supported kids by wielding her skills, experience, and proximity to them.

Keep in mind that pressuring employers to make greener changes at your workplace is an effective avenue. This can be as small as changing the toilet rolls to recycled paper or as big as signing the net zero emissions pledges for 2050 that many large companies have signed onto (e.g., Microsoft, Pepsico, and Apple). Unions can be supportive engines for pushing employers toward clean energy. In 2019, Amazon Employees for Climate Justice staged a multicity walkout, urging Amazon to reduce fossil fuel dependence. In response, CEO Jeff Bezos announced a climate pledge of reaching net zero by 2040.

Don't be a discounter! Get involved in legacy initiatives for future generations. I've found tremendous inspiration in the book *The Good Ancestor: A Radical Prescription for Long-Term Thinking* by philosopher Roman Krznaric, which advocates for holding a long-term perspective and finds ways to keep future generations in mind. Krznaric uses the term "discounting" to describe how policymakers fail to adequately factor in harm to future, unborn generations. Here are a few inspiring examples of legacy initiatives (the first two of which are mentioned in Krznaric's book):

- Future Library is a project where one thousand trees were planted outside of Oslo, Norway. These trees will be cut down and pulped

over the next century to create one hundred limited-edition anthologies by invited authors as a gift to future generations.
- Svalbard Global Seed Vault was created inside of a rock bunker in the Arctic to preserve a million seeds from six thousand species for at least one thousand years.
- Japan's Future Design movement is a model of participatory decision-making that balances the needs and concerns of current residents with those of future residents through an experiential simulation.
- Dear Tomorrow is an online video project providing a forum for people to communicate with future and younger generations. It is a space meant for sharing cross-generational hopes, dreams, and sentiments.
- Climate Stories Project curates personal stories about personal and community responses to climate change to be shared across the world, and includes different generational perspectives.

These legacy initiatives are intergenerational by design. Not all of the founders will see the fruits of their labors—and that's the point!

Empower through environmental education. Environmental education is a way for adults to teach youth about how the natural environment functions and consider ways that humans can live sustainably, instill values, and prepare a new generation of climate leaders. There are plenty of examples and innovations in this field. One inspiring local example is a Berkeley science teacher project to engage elementary school–age kids in local reforestation. Neelam Patil studied and taught a method for planting trees into fast-growth forests attributed to Japanese botanist Akira Miyawaki. In 2021, she and other adults worked alongside fourth and fifth grade classes at two local elementary schools on this project, digging into soil and gaining hands-on experience planting saplings, which will grow into super-dense forests around their schools. Not only do projects like this foster connection to the MHW, but they also empower kids to participate in solutions, increasing a sense of hope, agency, and building community.

Another example is Xoli Fuyani, who is an environmental educator in Cape Town, South Africa. She runs several hands-on, interactive programs at schools in marginalized communities, focusing on cultivating a relationship between kids and the living world. For example, in her first class, she introduces each child to their own earthworm, whom they will care for. The children eventually grow food in a garden enriched by their earthworm friends. For many of her students who live in low-income urban environments, this experience is revelatory and exhilarating. Xoli understands environmental education is a fundamental step to inspiring new generations of youth activists and leads two youth empowerment groups: Eco-Warriors and Black Girls Rising.

Offer youth a seat at the table whenever possible. An intergenerational climate movement demands inclusivity, and this means giving youth a seat at the table. This principle is certainly embedded into many Indigenous communities, in which youth are seen as vital participants in decision-making processes. The American Climate Corps, launched by the Biden administration in 2024, centralized and empowered young people (though canceled by the Trump administration the following year, it's a model that can be reinstated under another administration). Transitioning toward environmental justice models in which marginalized communities are involved in decision-making processes also centralizes and empowers young people.

Years ago, as a high school student and copresident of our environmental club, I was invited to sit on a panel with British anthropologist Jane Goodall (her headquarters were then in our small town in Connecticut) as she had recently launched the Roots to Shoots program. In a local TV interview, I experienced a sense of pride and relief that counterbalanced my nerves that day as I voiced my point of view as a youth about environmental issues my group was interested in. Sometimes, in my small town, it felt as though I was screaming into the void, but this experience validated that what our group cared about mattered. Looking back on the experience now, I can see how it helped to shape me into who am I today: a climate-engaged anti-doomer adult. It has reinforced my belief that we must amplify youth voices and create cross-generational connections whenever possible, and be sure to include marginalized voices and frontline communities as well.

Be open to relearning. Intergenerational gaps naturally occur as our culture evolves over time. However, your willingness to stay as current as possible around youth culture—whether through language, concepts, or ideas—and finding ways to acknowledge and mindfully approach your degree of power and privilege will help smooth out relations. Young clients in my office often school me about new terms and slang. Sometimes they'll see a look of bewilderment cross my face, and they'll ask me, "Do you know what ___ is?" As much as I always wish I was that cool, I have to own up that I don't know, and then they'll quite happily explain. We must be continually willing to relearn and stay humble in order to maintain a healthy connection to our young people; that's the ticket to fuel intergenerational allyship.

Consider giving your child your vote. In his 2020 TED Talk "How to Be a Good Ancestor," Roman Krznaric shares a decision that he made with his wife to gift their votes to their twelve-year-old twin boys during elections in England.[5] Children stand to inherit the choices made today, he reasons, so they should cast the votes. The four of them debated politics at the dinner table before parents cast votes on behalf of their kids' choices. While this act may seem radical, it makes logical sense and fosters the intergenerational communication and trust we desperately need at this time.

Celebrate Earth Day. I have a dream that one day soon our culture (or better yet, our global culture) will declare a shared vision for environmental values, and in doing so, instill a greater sense of hope and congruency within our psyches. In a 2024 article in the *San Francisco Chronicle*, I noted how a federal holiday would bring environmental stewardship and awareness into public consciousness. I don't see this as radical, but rather as practical. But beyond that, it's a vital signal to young people—many of whom are struggling immensely as a result of the inability of adults to make progress on this vital issue—that their elders are willing to tackle our climate emergency as an intergenerational society.

Storytelling. Stories are endemic to our very survival. They have potency and power, can shake us out of stuck places, and emotionally arrest us when we least expect it. Stories revitalize our spirits, allow us to raise our collective voices, help us share critical knowledge and our visions for the world, and unleash courage and vulnerability (two sides of the same coin). Think about how transfixed you might be listening to an episode of *The Moth*, or your kids are listening to *Circle Round*, or some other podcast.

But just as stories are a potent catalyst for change, they can also hold us back. As parents, we sometimes impose our own stories onto our children without even realizing it. I worked with a father whose childhood experiences influenced his perspective on how his daughter grew up, to the point that he had some very high standards for her. The pressure he exerted on her to be successful drove a wedge into their relationship that he began to understand only by examining his own wound baggage. But the story had become entrenched over the years, and there was a reluctance for him in letting it go.

Narrative therapy focuses on exceptions to our stories, or those aspects of stories that don't align with the main thrust of the story and may allow us an alternative channel to see things differently. Exceptions can be an avenue for breaking out of habitual patterns. As parents, we must remember that even as we make space to honor stories, we mustn't cling to them or stifle others with them. We all need space to grow, to emerge, and to flourish.

The beautiful thing about stories is that they're constantly evolving: people change, contexts change, and perspectives change. The climate crisis is a good example of an evolving narrative. Joanna Macy has wisely articulated that multiple stories are playing out at the same time in the climate narrative: business as usual, The Great Unraveling, and The Great Turning. And the ending is not yet written. That idea should breed hope!

Recalibrating the Silent Dance Party

As it turns out, there are many avenues for intergenerational work—it's quite accessible. Parents are naturally poised for intergenerational work as they interface with younger generations all the time. Often the hardest part is getting started, and once you do, the rest falls into place. As you choose your own adventure from the sampling of opportunities discussed in this chapter, be sure to hold compassion for yourself and gratitude for your willingness to try.

By bridging together generational differences and meeting rifts with compassion and empathy, our silent dance party will transform into a unified choreography. Out of disjointed, competing channels will come one channel, one rhythm, one powerhouse collective body.

Dominican author Junot Diaz's children's book *Islandborn* captures the poignant potential of intergenerational transmission. The book tells the story of Lola, a young Dominican-American girl who's given a school assignment to draw a picture of the country where she's from. Since she came to the US as a baby, she has little memory of life in her homeland. She interviews family and community members about what life was like on "the island" and receives a range of sensory-inspired responses: music and dancing in the streets, "color everywhere," beautiful beaches, and "dolphins like poetry." But it's her elder landlord, Mr. Mir, who finds a respectful way to tell Lola the story of the "monster"—dictator Rafael Trujillo—who ruled the island with crushing power and violence.

Ultimately, Lola learns how "strong smart young women . . . and men" rose up as heroes to defeat the monster. Not only does she integrate tragedy into her understanding of her cultural history through her class assignment, but she also learns that her own grandma was a hero in the resistance movement. Lola's memory, identity, and growth are strengthened by intergenerational collective storytelling.

Like Mr. Mir, elders can be truth-tellers, possessing the courage and foresight to impart wisdom to younger generations. Intergenerational connections are powerful, and not just in the face of human dictators. In the face of complex and far-reaching demons like the one we are facing now, we need all hands on deck to joined together to fight this fight. Or, as I overheard at the coffee and cookie table at the NorCal Climate Elders retreat: "We need a level of collaboration we've never experienced before."

CREATIVE EXERCISE

UNITING HEARTS AND HANDS GROUP MURAL

Materials:
Paper, scissors, and colored pencils/crayons/markers/oil pastels
Option A: Poster board, glue, optional printed image of Earth/globe to be glued onto the poster board
Option B: Yarn, cord, or string, and clothespins or tape

Setup:
You can do this project as a family, an intergenerational group or community, or school class. It can be helpful to identify a group facilitator or a leader or two.

Prompt:
Each person traces one of their hands onto a sheet of paper. On another sheet of paper, they draw a heart of any size. Be sure to write your name and initials, and possibly your age, on the back of these items.

For the heart: Using shapes, colors and lines, or words, show what you appreciate or are grateful for about Mother Earth.

For the hands: Write or draw one way that you take care of, protect, or consider Mother Earth (or something that you plan to do).

Cut these out.

You can affix these onto a poster mural for option A. Or, for option B, hang, tape, or clip the hearts and hands from a string and hang this on a wall, across a room, or between two trees, or some other spot determined by the group. Carefully consider placement of the hearts and hands. Sometimes it can feel containing to have the hands framing the hearts.

Process:
As you create, notice without judgment any feelings, thoughts, and sensations arising in your body.

Product:
When you're done, step back and reflect on your creation. Perhaps journal and then share in dyads about your creations. After the mural is constructed, spend some time taking in the whole of it and sharing about the project as a whole with the group.

Reflection questions
- Did any emotions arise as you created your heart and hand?
- How was the process of making the mural and putting it together? Any challenges?
- What did you notice when you looked at the mural as a whole? Any intergenerational themes?

TEN

A New Vision of the Future

What Happens When We Sweat the Big Stuff, Do Right by Our Kids, and Stay Committed

> *"When I dare to be powerful, to use my strength in the service of my vision, then it becomes less and less important whether I am afraid."*
> —Audre Lorde

> *"Once we start to act, hope is everywhere. So instead of looking for hope, look for action. Then, and only then hope will come."*
> —Greta Thunberg

I'd like to kick off this last chapter by acknowledging you, the reader, for your stamina in making it to this point in the book. I know how hard it can be to finish things, particularly as a parent with constant disruptions!

Just imagine that I'm there with you, patting you on the back, high-fiving, fist-bumping, or dapping you. Maybe even giving you a good ole hug.

As you might know, endings aren't really endings at all. They are also beginnings. As they say: When one door closes, another door opens. While I'm sad to say goodbye, I'm also excited for you to go forth and do your thang! I'm grateful for your time and attention, and hope to cross paths with you some day. I'm not promising it will be as a professional, because it just as easily could be standing on line at Trader Joe's buying pancake mix or cozied up on a packed subway or BART car with our families.

Intertwined Fates in the "McWorld"

As we humans have continued to alter Earth's weather systems and cyclical processes, Earth in turn reminds us of our intertwined fates. The wildfires that scourge Pacific coastal forests and communities blanket the sky with smoke and particulates that blow eastward into other states, skies, and lungs. The tropical storms off the shores of Japan or Australia move eastward over the Pacific, hitting cold pockets at higher altitudes. They then trigger atmospheric rivers in California, resulting in unprecedented rainfall, snow, and flooding. Droughts or floods in agricultural lands are causing low-to-no yields, disrupting our world's food supply. And so on. My air is your air, and your backyard is my backyard.

Back in 1992, political theorist Benjamin Barber wrote an article in *The Atlantic* called "Jihad vs. McWorld." It described our intertwined fates amid an unfolding epoch of globalization as a "McWorld" tied together by technology, ecology, communications, and commerce. He wrote: "The planet is falling precipitantly apart AND coming reluctantly together at the very same moment."[1]

According to Barber, our world is characterized by an unsettling dialectic, a pushing and pulling apart as we simultaneously forge new ways of communicating and relating. In the process, we move toward global homogenization amid technological distances and loneliness. The question we are grappling with in all of this mayhem is how in the heck do we raise our kids amid all of this chaos and instability?

Indeed, humans have created a fast world, one that romanticizes technology, efficiency, and wealth accumulation at the expense of so many other things that many of us cherish: peace, family time, sparking joy, playing, spending time in nature, reading, quiet moments, and even mindful labor. Our ancestors and our leaders have helped make the bed that we lie in today, and many of us continue to make it a part of our MA.

No country or culture can unmake this bed unilaterally. It will take many multigenerational villages (read: governments, international bodies, NGOs, businesses, communities) coordinating efforts, or at least working toward a similar goal. On a personal level, it will also take the softening of our hearts, and for those with privilege, a relinquishing aspect of our untenable lifestyles. Through all of this change we must not lose sight of the moving dial within us all: hope.

Hope Is Not a Destination; It's a Compass

In the climate world, "hope" is a frequent subject of debate. How much should we actually have? Is it helpful or harmful? Isn't attaching to a desired outcome just a setup for failure? Norwegian psychologist Per Espen Stoknes describes four possible versions of hope, and I find his framing very useful[2]:

1. Passive Hope: aka Pollyanna hope
2. Heroic Hope: We can fight and make it happen, aka hope on steroids
3. Stoic Hope: A grin-and-bear-it type of hope, but with little action or intention
4. Grounded Hope: "Grounded in our being, in our character, and calling, not in some expected outcome"

There's a tendency to assume that when *you* think of hope and *I* think of hope we are thinking about the same thing. But that's not always the case. There are many variants of hope. Here are two more hot takes while we're at it:

- *Active hope.* Joanna Macy and Chris Johnstone write about the idea that we must remain actively engaged in realizing the change we wish to see in the world. Active hope is both a practice and a process, like meditation or gardening. No matter what's happening around us, we allow our intention to be our guide.
- *Radical imagination.* Climate activist Tori Tsui talks about our ability to imagine despite the oppressive systems that stifle imagination and curiosity. Radical imagination builds what she refers to as "stubborn optimism," a hope that things will get better. Often, Tsui says, those with lived experiences of crisis and oppression tap into a form of stubborn optimism in order to persevere and even survive.

For me, what feels important to impart here is that hope is a moving dial, like a compass, shifting in response to both external stimuli (media, local climate impacts, conversations) and your own internal conditions (based on how resourced you feel, your mood, other stressors). We can't possibly predict the future and whether it will be what we are hoping for, but we can trust hope as our internal compass if we're headed off course. Hope is adaptive, and it can help us to adjust in order to move around obstacles that crop up.

Hope is also distinct from optimism. Optimism is a passive belief that there will be positive outcomes in the future; hope is an emotion linked to a sense of action and agency. Springing out of that action and agency is a wellspring of power and motivation that can help to chart a path forward. We can nurture hope communally, as embers that we take turns stoking. Is there an area of hope you would like to cultivate?

Author and activist Rebecca Solnit wrote, "Hope is not a lottery ticket you can sit on the sofa and clutch, feeling lucky. It is an axe you break down doors with in an emergency."[3] Hope means acting, intending, imagining, and seeing with clear eyes and a full heart. It's holding the twin beliefs that a brighter future is possible *and* that you have some power in making it brighter. It's an act of survival, and we must keep finding our way toward it.

How to Steer Clear of Doomerism

In the climate world, there are many factors that stir up a Climate Anxiety Cocktail: doomsday media headlines and articles; grief-stricken, depressed, or panicked folks who we come into contact with; characters in movies, books, streaming media, and podcasts. Sometimes you feel smothered by a downward spiral of negativity. Your internal response reflects the devastation that you see happening around you, and that becomes self-reinforcing, limiting your actions and behaviors, even changing your perspective about the fate of humanity. That's not necessarily doomerism, however. That's being a human who's paying attention.

As I see it, the two main differences between processing deep climate emotions and succumbing to doomerism lie in the following:

(1) The longevity of negative climate emotions that you experience (e.g., lasting for years rather than weeks or days);

and

(2) Identifying with a cynical, nihilistic, or fatalistic worldview.

Doomers become rooted in their cynical perspectives to the point that they cling to it as their identity. Further, they might become more focused on not budging and rationalizing their perspective than on evolving. In that way, Doomers are victims of their own identities.

It's worth noting that when a person faces a threat of such magnitude that they perceive a lack of control, they may avoid engaging at all as a form of self-protection. In nervous system language, they may enter a "freeze" state. Doomer-heavy media blitzes such as apocalyptic or defeating headlines can paralyze folks and reinforce doomerism. Even despite positive media intentions to shine a light on these issues, saturation in doomer-speak creates a vicious cycle of ecoparalysis.

Doomerism does us no favors. Like a toxic algae bloom, it keeps growing over time, feeding itself, and proliferating. If we don't want to spin into a downward mental spiral that will paralyze us and inhibit action, then we must, must, must maintain a relationship with hope and starve this toxic bloom by not consciously feeding it. And yet, it's important to offer the reminder that our internal self-protective mechanisms—of fight, flight, and freeze—can make this very difficult because these are responses based in the physiology of perceived threat. Let's take a moment to be curious about the Doomer perspective, as building awareness is one key toward freedom.

I see three categories of people in the doomerism realm: doomers, gloomers, and preppers. Doomers are hell-bent, Scrooge-like naysayers that scoff and say "Bah! That's impossible!" or "Never gonna happen!" As we might tell our kids, "Well, it won't with that attitude!" There is a human impulse to cut ourselves off at the knees by preparing for disappointment. But it's incredibly self-defeating to think and speak aloud negatively ad nauseum about our prospects. It makes for a self-fulfilling prophecy, or at least self-sabotage. This kind of negative hot air, or self-talk, doesn't capture the full potential of our human nature; it keeps us locked into a vicious, self-and-planetary-defeating cycle.

While doomers are actively vocal in their curmudgeonliness, gloomers are more passive. Gloomers quietly suck away all the oxygen that we need to energize our climate movements. Much like *Winnie-the-Pooh*'s Eeyore, they might groan, bemoan, and say things like "What's the point?" or "The planet's going to shit, accept it already." They are a bit like the anchor that keeps the ship from ever leaving the harbor. How's that for a climate engagement strategy?

And then there are the preppers, whose focus is on individual preparation for the anticipated apocalypse. I'm not talking about people who are preparing for extreme weather events and stocking up on canned goods, flashlights, and water. Nor do I mean those brushing up on DIY skills like first aid, carpentry, or wilderness survival. Those types of cautionary measures make good sense and can help out communities. Preppers, in my definition, are way more extreme. They are the folks whose imaginations have run wild and are consciously or unconsciously disregarding the well-being of others by putting themselves first. Billionaires with outsized resources, such as Elon Musk, pour precious money into cryogenic freezing operations and space colonization with only themselves (and their coteries) in mind. Mark Zuckerberg is building a five-thousand-square-foot underground bunker in Hawaii. And the list goes on . . . me, me, me is the name of the game.

Preppers also create self-fulfilling prophecies because they make choices with real impacts that siphon off energy, talent, skills, and funds that could be better invested in advancing climate measures, education, and solutions than in their efforts to save themselves and their inner circle. They are an ugly offshoot of the egomania and greed that got us into our current debacle in the first place. The reality is that we need cooperative, prosocial behaviors that will strengthen our response and build momentum, shape policy, and pressure the private sector.

Doomerism is not a life prescription. You can make a conscious effort to see things differently. You can start by reorienting your inner compass to hope by spending time with anti-doomers, reading positive news stories, reshaping negative narratives, and partaking in the toolkit in chapter 6.

As singer-songwriter Jose Gonzales sings: "Don't let the darkness eat you up."

Raising Anti-Doomers

One night, after my then four-year-old daughter and I had returned home from an event at the local Jewish community center, she broke into song—but not one that gave me the warm fuzzies (like her spontaneous musical performances usually did). She was repeating lines written to accompany Alicia Keys' "Girl on Fire" that she'd overheard at an adult sketch comedy rehearsal. Swaying gently, her golden tresses a halo around her head, she sang in near perfect intonation: "This *world* is on fire . . . this world is on fire . . . " All I could do was gape in horror, while my heart plunked into my stomach.

To be a parent in this day and age—or in any day and age—is undoubtedly a blessing. But the job description now includes, among many other things, worrying about ebbing biodiversity, fatal mass shootings, and forced migrations.

Raising anti-doomers in this world calls on us to be adaptive, responsive, and compassionate adults who are willing both to listen and validate our kids and to share our own thoughts and feelings appropriately. Raising anti-doomers requires a mutual respect between parents and their vulnerable offspring.

If we let doomerism fester inside of our homes, then we lose the battle of human agency. As parents, we are holders of hope just as much as our kids. Leaning into hope is something that we all must do now; our collective well of hope will grow the more we all lean in.

At the same time, slipping into what's called "hopium"—clinging to false hope because it's hard to accept the truth of what's happening—is problematic. You might think something like: "Things have a way of working themselves out—it will be okay in the end" or even "Gen Z is super-smart and they'll save the world." These thoughts let us off the hook, breeding complacency and disavowal.

Hopium sedates, and doomerism suffocates. Both are examples of black and white thinking, which often surfaces during stressful periods in your life, or in human history. But black and white thinking inhibits engagement; neither pole is skillful.

To help check this tendency toward extremes, I've developed a visual that I call the "Anti-Doomer Sweet Spot," the space parents need to occupy to raise kids who are empowered to tackle the obstacles that lie ahead. The "Anti-Doomer Sweet Spot" illustration shows the range of other possible responses that exist between the poles of hopium and doomerism—all of them more helpful and adaptive!

What you see in the illustration is a grey scale of many of the different responses we might have to the climate emergency. It is meant to breathe nuance into the doomer-hopium chokehold. This ecosystem shows the diverse range of responses that span the poles between hopium (the balloon floating skyward) and doomerism (the deadweight anchor on the sea floor). Looked at in this way, we can see just how extreme those two

A New Vision of the Future

responses are, and we can get to know some of the other creatures—and gradations—inhabiting this nutrient-rich ecosystem.

Fear and guilt are the ducks skimming the top of the water, showing how our emotions are really opportunities (and need not be sitting ducks, so to speak). But if we succumb to them in unhealthy ways so that they burden, harden, or polarize us, then we sink down or fly away.

I spoke about various strains of hope earlier in the chapter, but there are a few variants appearing in this graphic that I haven't yet addressed.

> **Stubborn optimism:** Costa Rican diplomat Christiana Figueres has defined this as "the mindset that is necessary to transform the reality we're given into the reality that we want."[4]
> **Cautious optimism:** This is optimism tinged with realism so that you're prepared for any outcome.
> **Constructive hope:** The ability to face risks and uncertainty while maintaining positive beliefs and actions.[5]

There are some other concepts in the Anti-Doomer Sweet Spot worth mentioning as well:

> **Personal greenwashing:** The conflicted or ambivalent eco-behaviors that fall short of true engagement.
> **Realistic pessimism:** Not trusting in a positive outcome but not succumbing to doom either; trying to stay balanced.

In a way, we're in parallel process with our declining ecosystems; as biodiversity wanes, so has our imagination succumbed to doomer tendencies devoid of hope and imagination. We must make a conscious effort to reclaim human imagination to solve this crisis. Seeing reality in shades and gradations, and reclaiming nuance, rather than seeing it in rigid polarities, is vital.

As an artist, a writer, a therapist, and human being, I appreciate the value of "grey-dations." Notably, emotions and memories swirl in the grey.

In this book, I spend a lot of time fleshing out choices, alternatives, and flexibility so that we can expand our selves into our fullest potential. If we stay contracted, then we are small, limited, and helpless.

There's a town I like to visit along the Northern California coast that illustrates this point beautifully. It's nestled into an estuary where the freshwater river empties out into cold salty Pacific waters. While kayaking there for the first time, I floated along a calm, quiet river channel inland, and my boat bobbed along over clear, fresh water. But as I approached the mouth of the ocean, the water gradually turned murkier and became choppy as fresh and salt waters comingled. The ecosystem had changed at this nexus. Pretty soon, the sound of powerful, epic waves crashing just beyond the point sliced into the quiet calm. At this meeting ground, I froze in an awe laced with glee, allowing the kayak to rock me; I could feel the power of straddling two connected worlds. I could feel both the vulnerability of the ecosystem, and my own, rocking in the boat.

Sweet spots are tender places and good reminders of the connection points that exist between any polarities: good and bad, happy and sad, black and white, light and dark. If we avoid feeling vulnerable, we miss so much of our emotional experience, and the opportunity for gaining perspective. More urgently, we miss staying awake and alert to the crisis at hand.

Solution-Focused Thinking

Buddhist teacher Joan Halifax coined the phrase "strong back, soft front" as a way to hold conflicting dualities; the toughness and strength required during difficult times must be balanced by openness and compassion. What I love about this metaphor, in addition to naming another sweet spot, is that it's a reminder that we can feel into this in an embodied way.

PARENT PAUSE

Go ahead and try out this micro-adjustment and see what you notice. Elongate your spine by rolling your shoulders back, take a breath in and out, while feeling openness and expansion across your heart region. Take another deep breath in, feeling into the strength of your back. As you exhale slowly, feel into a gentle release across your front. This is a nice way to toggle and reconnect with the grey (and you may just activate your vagus nerve while you're at it). Keep visiting this sweet spot as often as you can.

Another way to get moving in the direction of hope along the Anti-Doomer Sweet Spot is by focusing on positive stories and solutions. Try these on for size: The number of youth-led climate change lawsuits is fast growing and making inroads, such as the wins in Montana and Hawaii mentioned earlier. Congress's Inflation Reduction Act of 2022 was the largest investment ever in clean energy, driving up manufacturing in the solar and EV sectors—and it can be reinstated in the next presidency. The Twenty-Eighth Conference of the Parties finally established a loss and damage fund for low- and middle-income countries dealing with the harshest climate change impacts. California, as the world's fifth largest economy, continues its aggressive climate commitments to date, including the California Global Solutions Act of 2006 aimed at reducing greenhouse gas emissions, investing in clean energy, and the phasing out the sales of new gas-powered cars by 2035.

The truth is that we have plenty of Indigenous-led solutions, nature-based solutions, and scientific technologies already that can help curb, draw down, and sequester emissions, and regenerate the Earth. Part of reinforcing anti-doomerism is spending time with all of the innovations and regenerative

Indigenous practices that exist, and that are being tested or implemented in the climate mitigation and adaptation realms. Please check out the appendices for resources to learn more about ocean draw down, the seventh-generation principle, and the school bus electrification movement.

Psychoanalyst Sally Weintrobe calls this a shift from Planet LaLa to Planet Life. This shift entails challenging the economic notion of endless growth by respecting Earth's finite resources and boundaries through a commitment to sustainability. And instead of fruitless debating whether or not the answer is all in degrowth (reducing global consumption and readjusting our methods of determining GDP) versus green growth (scaling up green innovations that upholds our current habits and lifestyles), it's another both/and response. Once again, these solutions should not be viewed as dialectics, but rather in terms of overlap, and sweetspots, and how to just keep moving forward on both fronts.

Biomimicry, the practice of learning from and mimicking strategies found in nature to solve human problems, offers us both solutions and hope as we navigate these tumultuous times. Moving out of an extractive toward a more harmonious relationship with the MHW to create more sustainable life systems is the change that we so desperately need—and can be a part of.

In my twenty plus years of living in Northern California, I've been entranced by redwood trees, which dominate the coastal landscapes here. A remarkable characteristic of redwood trees infuses me with hope. I offer this an example of biomimicry. When a parent tree dies—be it from an axe, wildfire, or other source—sprouts from its root system shoot up in a circle around the parent tree, forming what is called a fairy ring. This forms the basis for a new generation of trees that utilize the root system and nutrients of the parent tree to grow. There is a poignant parent-child metaphor in this, of course. But more than that, I think there's a way we can be reminded that out of hardship and destruction there is a chance for rebirth and renewal. Our allies in the MHW are constantly reminding us of hope in this way.

Scaling Resilience from the Ground Up

It's time to circle back to you, dear Parent Club, and appeal to your internal locus of control once more. Some readers might be hoping for Big Change to come from the Top. Indeed, let's hope it is. But simultaneously, we must keep pushing individual and family action forward—not only because it's helpful for our mental health but also because it will help seed a groundswell movement to hold corporate and government entities responsible for climate action.

Not only do small actions make a difference in the equation, they also mirror the larger cultural shift that we need to see toward empathy, cooperation, and responsibility. Taking action lessens individual anxiety and encourages other individuals to do the same. If a parent catches wind of another parent's efforts, like recycling electronics or buying an electric vehicle, studies show that they might feel compelled to do that, too. And let's not forget that living out of alignment with your values confuses kids and can contribute to their sense of hopelessness and despair. Creating an "internal locus of control," a sense of internal agency, is a powerful tool when so much is happening outside of our direct control. Please keep in mind as this book wraps up: *You* are part of the solution!

When I was a teenager finding my activist legs, I happened to pick up a book at a local book fair that included the story "The Star Thrower" by American anthropologist Loren Eisley. The story is about a man who's walking along the beach one day, when he notices a boy picking up starfish that have been left behind by the ebbing tide. He watches as the boy, one by one, throws starfish back out to sea. When the old man asks him about it, the boy says that they will die if he does not do this. The old man finds this amusing and points out that there are too many starfish for him to possibly save, so his doing this won't make much difference. After listening politely to the man, the boy bends down, picks up another starfish and throws it out to sea, saying: "It made a difference to that one."

I can't think of a better place to end this book. Adults, in all their life-accrued wisdom, often overlook simple solutions and possibilities. But if we, the Parent Club, pay attention, children will be our mirrors, reflecting back all that we need to see to make better choices for the future. Once upon a time, optimism bloomed inside of our own youthful hearts. Returning to that place of deep empathy, love, yearning, and care will spur continued authentic engagement.

Your Climate Manifesto–and Beyond

By now, I hope that you feel different in some way: perhaps calmer, perhaps clearer or more resolute in your commitment, perhaps more confident to show up around your kids and in your community.

As you trust the emergence of your parent identity and where it leads you, your journey will unfold. Picture this: parents across the world holding up lanterns into darkness. One light is a trifle, but a thousand lights brighten up the surroundings, and a million—or more!—flood and transform the darkness into a connected band of light extending the full circumference of the globe. Each hand that holds a lantern has a story, a dream for a child—laughing, playing, healthy, and free. These hands are tender yet strong, and they shine light in the face of encroaching darkness. Each hand helps to protect another's flames. That is the full might and potential of the Parent Club: to keep the light in this world alive and to never, ever, lose hope.

A final multiple-choice question for you.

After reading this book, I will:

a. Go stick my head back into the sand
b. Make small changes in my life/family/ behaviors
c. Make big or radical changes in my life/family/behaviors
d. Invite others into the climate Parent Club
e. Spread the message of anti-doomerism with fervor in my home and community
f. Other: _____

Raising Anti-Doomers

PARENT CLUB MANIFESTO

I,_____, HEREBY ASSUME AN ENGAGED AND/OR PROACTIVE ROLE IN THE _____ AFFECTING US/MANY OF US TODAY.
(CRISIS OR CAUSE OF YOUR CHOICE)

I DO THIS NOT ONLY FOR MY KIDS, BUT FOR THE WELLBEING OF CHILDREN EVERYWHERE IN CONSIDERATION OF THE NEXT SEVEN GENERATIONS.

THE CORE VALUES I HOLD WHICH CAN FUEL MY ENGAGEMENT IN THIS CAUSE ARE: _____

IN THE PAST, I HAVE BEEN _____ x INVOLVED IN THIS CRISIS/CAUSE. GOING FORWARD, I COMMIT TO BE _____ x INVOLVED. I UNDERSTAND THAT MY LEVEL OF ENGAGEMENT MUST BE HEALTHY AND SUSTAINABLE FOR ME. I PROMISE TO READJUST IT AS NEEDED, AND WON'T BEAT MYSELF UP IF I FALL DOWN ALONG THE "YELLOW BRICK ROAD." I VOW TO RETURN AGAIN AND AGAIN TO THIS PATH.

I PROMISE TO REFRESH MYSELF WITH MY CUSTOMIZED ANXIETY TONIC, AND CONTINUE WITH MY PARENT CENTERING PRACTICES (PCPS) SUCH AS _____, _____ AND _____.

I WILL DO EVERYTHING IN MY POWER NOT TO SUCCUMB TO DOOMERISM OR HOPIUM BY MAINTAINING SELF-AWARENESS AND LEANING ON MY COMMUNITY FOR SUPPORT. I WILL ENERGIZE THE INTERGENERATIONAL ANTI-DOOMER MOVEMENT AS MUCH AS I CAN.

THE PARENT IDENTITY, OR ROLE, THAT I ENVISION FOR MYSELF IS_____. SOME OF MY PROJECTS WILL INCLUDE: _____. IN ORDER TO GET STARTED, HERE'S WHAT I NEED TO DO: _____, _____ AND _____.

MY TIME COMMITMENT WILL BE: _____

MY ACCOUNTABILITY PARTNER OR GROUP IS: _____.

I WILL KNOW THAT I'M IN ALIGNMENT WITH MY VALUES, AND SHOWING UP AS BEST AS I CAN IF:_____
IF I "RELAPSE," HERE'S HOW I WILL KNOW: _____
I CAN GET BACK ON TRACK BY: _____

IF I FEEL VULNERABLE, THESE WORDS OF SELF-COMPASSION ARE HELPFUL: _____

THESE WORDS WILL HELP TO KEEP ME MOTIVATED: _____
(QUOTE/SAYING/PERSONAL MANTRA)

I UNDERSTAND THAT THE PARENT CLUB IS A RESOURCE THAT WILL HELP SUPPORT ME AND SUSTAIN MY WORK.

I SIGN THIS PARENT CLUB MANIFESTO, WITH AN OPEN AND FULL HEART, ON THE _____ DAY OF _____ IN THE YEAR _____.

MY HOPE FOR THE FUTURE IS THIS:_____

SIGNED,
_____
NAME

_____
NEW PARENT IDENTITY TITLE OR ROLE

APPENDIX A

Sample Automatic Negative Thought (ANT) Log

Situation or Event	Thought	Emotion	Behavior or Action	Compassionate Reframe

APPENDIX B

Self-Care Weekly Tracker

Activities/ Practices	Sun	Mon	Tues	Wed	Thurs	Fri	Sat

In the left column, list any activities or practices that you do—or aspire to do—that help to sustain your well-being. Aim for one to three activities per day. You can update the list at any time. Be sure to start each week with a fresh copy of this Tracker, and complete as you go or just review it at the end of the week to see how you are doing. Some examples of activities or practices include showering, walking or running, spending time with a pet or loved one, or attending a yoga class. Track anything big or small that helps you to feel a bit better, a bit more grounded, and well. You can add to this list while reading this book and include exercises from the Climate Tonic or the self-care acronym.

APPENDIX C

Tools for Managing Anxious or Traumatized Nervous Systems

- Guided meditations using an app (e.g., Calm, Insight Timer, Headspace), a timer, or soft music
- Water: swimming, cold plunge, warm shower or bath, waterfalls, or boating
- Walking, swinging, dancing, moving, martial arts, or exercise
- Forest bathing, nature lounging, sunbathing, or bare feet in the soil or grass
- Stretching, yoga
- Listening to music that intuitively feels good
- Playing an instrument, singing, humming, whistling, or growling
- Coloring books, crafts, journaling, painting, or eco-art
- Drinking a cup of herbal tea, inhaling the scent, then exhaling and blowing the tea to cool it. Repeat this calming cycle again and again. Bonus: Use fresh herbs if you have access or time (e.g., wild mint, sage, lemon verbena, or loose tea from a shop).
- Hand or foot massage with lotion, or family massage.
- Polyvagal exercises. positivepsychology.com/polyvagal-theory/

Appendix C

- Trauma Response Exercises (TRE); You can find a coach for this on the website. traumaprevention.com/
- Calm cards (a card deck of mindfulness exercises)
- Plum Village, "The Art of Mindful Living." Retreats and free meditations. plumvillage.org/#filter=.region-na
- Tara Brach's "RAIN" and other meditations. www.tarabrach.com/rain/
- The DailyOM. www.dailyom.com/
- Window of Tolerance. mentalhealthcenterkids.com/blogs/articles/window-of-tolerance
- Zones of Regulation. zonesofregulation.com/what-are-the-four-zones-of-regulation/
- *River of Life* (video). www.youtube.com/watch?v=ZVEDueyZ2C4
- "Owls, Watchdogs, and Possums, Oh My!" (Robbyn Gobbel). robyngobbel.com/wp-content/uploads/2023/08/Owls-Watchdogs-and-Possoms-Infographic.pdf

APPENDIX D

Indigenous-Focused Resources

Note: Please remember to distinguish between cultural appreciation and cultural appropriation. This appendix is meant to foster learning, growth, and healing in our relationship to the land and Indigenous communities.

Examples of **Traditional Ecological Knowledge (TEK):**

- **Reciprocity:** The idea that natural resources are not ours to commandeer; there is a give and take, a mutual and reciprocal relationship between humans and the MHW.
- **Gratitude:** Viewing the world through a lens of abundance rather than a lens of scarcity, expressing gratitude, and offering intentional gratitude at the outset of an event or activity.
- **The Seventh Generation Principle:** A Haudenosaunee (Iroquois) concept that the decisions we make today will impact our descendants, and so we should live as though we are borrowing the Earth from the next seven generations.
- **Intergenerational Cooperation:** Prioritizing the well-being, involvement, and inclusion of all generations in decision-making spaces.

Appendix D

- **Regenerative Economy:** An economy based on ecological restoration, community protection, equity, justice, and fair participation (as framed by the Indigenous Environmental Network).
- **Gift Economy:** An exchange based on gift giving without expecting anything in return; this creates a positive debt in which people are tied to each other through relationship (as expressed through gratitude, reciprocity, and interdependence).

Resources
Books

Sand Talk by Tyson Yunkaporta
Braiding Sweetgrass: Indigenous Wisdom, Scientific Knowledge, and the Teachings of Plants by Robin Wall Kimmerer
Restoring the Kinship Worldview: Indigenous Voices Introduce 28 Precepts for Rebalancing Life on Planet Earth by Wahinkpe Topa and Darcia Narváez
What Kind of Ancestor Do You Want to Be? edited by John Hausdoerffer, Brooke Parry Hech, Melissa K. Nelson, and Katherine Kassouf Cummings
Red Alert! Saving the Planet with Indigenous Knowledge by Daniel R. Wildcat

Children's Books

We Are the Water Protectors by Carol Lindstrom and Michaela Goade (illus.)
The Great Kapok Tree by Cherry Lynne
Be a Good Ancestor by Leona Prince, Gabrielle Prince, and Carla Joseph (illus.)
Braiding Sweetgrass for Young Adults by Robin Wall Kimmerer, Monique Gray Smith, and Nicole Neidhardt (illus.)
Indigenous Ingenuity: A Celebration of Traditional North American Knowledge by Deidre Havrelock and Edward Kay

Documentaries

The Reciprocity Project
Mossville: When Great Trees Fall
The Candor and the Eagle
Remothering the Land
One with the Whale
Keepers of the Coast

Appendix D

Land Back: The Indigenous Fight to Reclaim Stolen Lands
Facing the Storm: The Indigenous Response

Non-profits

Indigenous Environmental Network, www.ienearth.org
Indigenous Climate Action, www.indigenousclimateaction.com
Indigenous Climate Hub, indigenousclimatehub.ca
Climate Atlas of Canada, climateatlas.ca

APPENDIX E

Resources for Connecting Low-Income and BIPOC Communities with the MHW

Every Kid Outdoors. Every fourth grade child qualifies for a free one-year pass to national parks for their families. everykidoutdoors.gov/index.htm

Children & Nature Network. Leading a global movement to increase equitable access to nature so that children—and natural places—can thrive. www.childrenandnature.org/

Gateway to the Great Outdoors. "[E]mpowers students from low-income households by igniting curiosity through outdoor experiences, building environmental science literacy, and promoting healthy lifestyles." www.gatewayoutdoors.org

The Outdoor Foundation's "Thrive Outside Initiative." Community-led initiatives that provide children and families with repeated and reinforcing experiences in the outdoors. outdoorindustry.org/participation/thrive-outside/

Anti-Racism in the Outdoors Guide. "Resources related to inclusion, diversity, equity, and access of Black, Indigenous and people of color in parks and green spaces." www.publicgardens.org/wp-content/uploads/2020/08/anti-racism-outdoors.pdf

YMCA. Their camps and programs nationwide offer financial assistance.

GirlVentures. Combines outdoor adventure with socio-emotional learning for adolescent girls in the San Francisco Bay Area. www.girlventures.org/

APPENDIX F

Climate and Mental Health Resources

Books

Active Hope: How to Face the Mess We're in with Unexpected Resilience and Creative Power by Joanna Macy and Chris Johnstone
All We Can Save: Truth, Courage, and Solutions for the Climate Crisis edited by Ayana Elizabeth Johnson and Katherine K. Wilkinson
Ancient Spirit Rising: Restoring Your Roots and Restoring Earth Community by Pegi Eyers
The Book of Awakening by Mark Nepo
Coming Back to Life: An Updated Guide to the Work That Reconnects by Joanna Macy
Daring Greatly by Brene Brown
Emergent Strategy: Shaping Change, Changing Worlds by adrienne maree brown
The Good Ancestor: A Radical Prescrption for Long-Term Thinking by Roman Krznaric
The Intersectional Environmentalist: How to Dismantle Systems of Oppression to Protect People + Planet by Leah Thomas
Last Child in the Woods: Saving Our Children from Nature-Deficit Disorder by Richard Louv
No One Is Too Small to Make a Difference by Greta Thunberg

Appendix F

Psychological Roots of the Climate Crisis: Neoliberal Exceptionalism and the Culture of Uncare by Sally Weintrobe

Books for Kids

All the Feelings Under the Sun by Leslie Davenport
I Am Jane Goodall by Brad Meltzer
The Lorax by Dr. Seuss
Rebel Girls Climate Warriors: 25 Tales of Women Who Protect the Earth by Rebel Girls and Christina Mittermeier
Rewild the World at Bedtime by Emily Hawkins

Documentaries

Choosing Earth
I Am Greta
Powerlands
Rockies Repeat
The Scale of Hope
This Changes Everything
Kiss the Ground
Not a documentary but . . . *Don't Look Up*

Podcasts and Newsletters

The Crucial Years, Bill McKibben
Unthinkable Times (formerly *Gen Dread*), Britt Wray
The Golden Hour: Climate, Children, Mental Health, Anya Kamenetz
Clear Dharma, Oren Jay Sofer
Talking Climate, Katherine Hayhoe
Climate Change and Happiness
Climate One
The Climate Question
Hot Take
How to Save a Planet
Outrage + Optimism: Climate Change

Climate Anti-Doomer Resources

Center for Regenerative Solutions, naturebasedclimate.solutions
The Daily Climate, www.dailyclimate.org/Good-News/

Appendix F

Grist, grist.org
Happy Eco News, happyeconews.com
Pachamama Alliance, pachamama.org
Project Drawdown, drawdown.org
Yes! Magazine, www.yesmagazine.org

Climate Creative Nexus

Artists for Climate, artistsforclimate.org
Climate Journal Project, www.climateculture.earth/directory/climate-journal-project
The Climate Museum, www.climatemuseum.org
The Climate Music Project, climatemusic.org
Climate Stories Project, www.climatestoriesproject.org
Hila the Kila (an eco-rapper and educator), hilatheearth.com

Climate Intergenerational Spaces

Climate Action Families, climateactionfamilies.org/about/
Climate Families NYC, www.climatefamiliesnyc.org
DearTomorrow project, www.deartomorrow.org
Dream in Green, dreamingreen.org
Future Design Council, www.souken.kochi-tech.ac.jp/seido/en/index.html
Talk Climate, talkclimate.org
Women's Earth Alliance, womensearthalliance.org

Climate Parent Groups

ClimateMama, climatemama.com
Madre, www.madre.org
Moms Clean Air Force, www.momscleanairforce.org
Mothers Out Front, www.mothersoutfront.org
Mothers Rebellion for Climate Justice, mothersrebellion.com
Mothers Rise Up, www.mothersriseup.org
Our Kids' Climate, ourkidsclimate.org
Parents For Future, parentsforfuture.org
Science Moms, sciencemoms.com
Sierra Club's Climate Parents, sierraclub.org/topics/climate-parents

Appendix F

Climate Psychology Resources

Climate Awakening, climateawakening.org
The Climate Aware Therapist Directory, www.climatepsychology.us/climate-therapists
Climate Café Network Hub, www.climate.cafe/
Climate Emotional Resilience Institute, climateemotionalresilience.org/
Climate and Your Mind, www.climateandyourmind.org
The Ecopsychepedia (online educational resource), ecopsychepedia.org
Good Grief Network, www.goodgriefnetwork.org/
The Work That Reconnects, workthatreconnects.org/

Please visit my website for some general resources:
www.ariellacookshonkoff.com

Acknowledgments

First and foremost, this book is for our children and the next seven generations. May they be permitted to feel the tingle of freedom in their bones, the calming comfort of human care and love, and their vital connection to all living things.

This book is also an offering in deep gratitude to Gaia, Mother Earth, Pachamama, and any and all conceptualizations of her. To that end, I'd like to humbly acknowledge the lands and waters that have raised me: Ramapo, Paugussett, Potatuck, Ute, Cheyenne, Arapaho, Mohican, Kanien'kehá:ka, and Ohlone lands. And I'd like to pay homage to my ancestral lineage that contributed to the formation of this book.

I'm a second-generation Polish-Romanian Jewish American on my mother's side. My grandmother Gloria Mark immigrated through Ellis Island in the early twentieth century. My grandfather Morris Mark immigrated before World War II. His parents were mercilessly killed in their village alongside countless others during the war. I have family members who survived Auschwitz in part because of their perseverance and resilience, as well as family members who perished in the concentration camps despite those same qualities. Their memories have shaped this work.

I'm a first generation British Jewish American on my father's side. The Cooks changed their name several times in England; it's a story of Jewish assimilation amid a backdrop of antisemitism. My father grew up in the

Acknowledgments

aftermath of World War II and used to sneak oranges as a child and eat them behind the couch—for they were considered a luxury.

After my brother Jonathan was born, my father, Philip Cook, and mother, Paula Cook, immigrated from the UK to the US. I was born in State College, Pennsylvania, five years later. My grandparents came from humble means—they were builders, bakers, and secretaries. They, and their offspring, were creative in their own ways: actors, dancers, musicians, writers, designers. My mother and father committed to creating a peaceful home, and achieved middle-class stability through entrepreneurship and editorial work. My brother served at USAID for a decade until the department was recently eliminated. These are only some of my ancestral stories, and I am grateful for how they have shaped me.

I want to thank the beautiful places from which I worked on this book over the past couple of years. California: Inverness, Gualala (St. Orres), Jenner, El Cerrito, Richmond Marina, Mount Shasta, and Mount Madonna; Costa Rica: Cahuita, Playa Hermosa, and Playa Portero; Mexico: Cabo Pulmo and Todos Santos; New York City; parts of Maine; western Massachusetts. Thank you to Lois Talkovsky, Susanne Leitner, and Robin Berrin who provided me with the gift of a space of my own to write in solitude.

First off, a warm cross-body hug of gratitude to my partner, Seth, who has been my emotional rock throughout this journey—holding down the fort when I needed to write; reminding me to celebrate the milestones along the way; and periodically grounding me in realism, science, strategy, and hugs.

A mama bear hug to my two open-hearted, tender, and fierce daughters, who are my truest teachers and joy mentors—reminding me to stay curious, to listen(!), and to play with them as much as possible.

To my parents who provided a relaxed, delightfully free, underscheduled home, a blessing for me that set up a life of creative inquiry, joy, and cultural exploration. The make-do, no-fuss, nonmaterialism in our household inspired travel, adventure, creativity, and intellectual pursuits.

Acknowledgments

A few shout-outs:

Mom: For your gentle patience, mastery in the telling of bedtime stories, love of word play, and frequent reminders to "look that up in the dictionary!"

Dad: For your playful spirit and joie de vivre entwined with ever-dogged perseverance, and your finding ways to make me laugh even in the hardest of moments.

Jon: For showing me the ropes of adulthood, expanding my "cool" horizons, and teaching me about the art of travel.

Aunt Zena: For being a model for me on how to be independent and self-reliant as a young woman.

Aunt Linda: For holding the family together like glue and practicing creative discipline.

My in-law clan (Heidi, Alan, Sam, and Laurel): For welcoming me into a vibrant, values-laden Jewish community that has fostered my growth in all directions.

Special thanks and love to my lifelong friend and platonic soulmate, Zoe Lewis, who is one of the most authentically real people that I know, a master validator, and who can make me laugh through tears and always makes me feel special.

Much gratitude to my wise and steady literary agent at Levine Greenberg Rostan, Lindsay Edgecombe, whose calm demeanor helped me to weather the various stages, bumps, and iterations of this sizable undertaking.

A chorus of thank yous to the Hachette team in working with me through unexpected transitions: to my first-leg editor Dan Ambrosio and my second-leg editor Renee Sedliar; to the glue that saw this book through from beginning to end, Nzinga Temu; and home stretch editors Allison Gudenau and Sean Moreau. Your visions for this book dovetailed and landed so beautifully in the end.

A warm thank you to my awesome illustrator, Pepita Sandwich, whose whimsical work enhances this book.

A special shout-out to my developmental editor Bridget Lyons, who helped to refine the ultra-raw material.

Acknowledgments

Thank you to my community of writer friends and cheerleaders: Lisa Romeo, Melissa Petro, Michael Lukas, Allison Landa, Oren Krell-Zeldin, Jenny Pritchett, Melanie Bishop, Leigh Marz, Sierra Millman, Dan Millman, the many voices in the Binders, etc.

Thank you to my second readers for finding time to read when time is such a precious resource: Kelsey Hudson, Wendy Greenspun, Dashal Moore, Deanna Jimenez, Rachelle Aiello, Josh Sonnenfeld, Jenny Miller, and Mor Keshet.

Thank you to those interviewed, both named and unnamed, here: Ella Suring, Maksim Batuyev, Carolyn McGrath, Robin Cooper, Lise van Susteren. Thank you to Climate Psychology Alliance North America's Executive Committee (particularly Barb Easterlin and Rebecca Weston) for granting me a writing sabbatical so that I could better focus on this project.

Thank you to my clients, both named and unnamed in this book, whom I continue to learn from each day as you evolve in your own unique ways.

And a *ginormous* hug to my local Parent Club, who have heard me yammering on like a broken record about writing getaways, deadlines, and the overall hecticness of life with raising two kids, seeing clients, and working on this book. I could name names, but you know you are! And a special mention to the incredibly talented and smartass women in my mama's groups.

This book integrates strands of wisdom gleaned in my forty-five years. But to call out a few teachers that have informed my work: Tara Brach, Derek Sheahan, Eugene Cash, Joanna Macy, Oren Jay Sofer, Dina Amsterdam, Ai Kubo, Gingi Allen, Carrie Katz, Nathalie Babazadeh, Patricia "Trix" Adler, Fei DaCosta, Julie Sigoloff, Julia Cameron, Brene Brown, Joseph Cornell, Mary Ann Eddowes, Kathryn Davis, Douglas Glover, Devera Jackson-Garber, Debbie Diaz, Kara Wahlin, Linda Chapman, Deann Acton, Laury Rappaport, Richard Carolan, Gwen Sanders, Brian Lukas, David Akullian.

Acknowledgments

Writing this book has infringed on my family life at times, exhausted me, pushed me to my limits, and exposed my vulnerabilities, but it's something that I've felt compelled to do—an offering that I knew I could make good on. Started on the heels of the pandemic and carried through a devastating long-lasting war in Gaza and Israel, through the 2024 elections, and 2025 inauguration, and even as the fallout and fear of a rapacious government continues, I wrote and I wrote. And on the Micro level, through the grief of a devastating family loss, bouts of COVID-19 and illness, and a major home renovation that sent us unexpectedly packing and adapting. My computer survived a number of ordeals, from freezing and crashing, to nearly coated in vomit in the car, to almost left behind at airport security. Just like parenthood, nothing ever stopped for a minute. Life continues.

And, finally, I rest. For now.

Notes

Introduction

1. Elizabeth Allured, interview with the author on November 12, 2020.
2. Intergovernmental Panel on Climate Change, "In-Depth Q&A: The IPCC's Sixth Assessment on How Climate Change Impacts the World," CarbonBrief.org, February 28, 2022, www.carbonbrief.org/in-depth-qa-the-ipccs-sixth-assessment-on-how-climate-change-impacts-the-world/.
3. Kim Nicholas, "Climate Change: 'We Can Fix It World Cafe' 2018," *kimnicholas.com* (blog), November 5, 2014, www.kimnicholas.com/we-can-fix-it-world-cafe.html.
4. António Guterres, "Secretary-General's Statement on the IPCC Working Group 1 Report on the Physical Science Basis of the Sixth Assessment," United Nations, August 9, 2021, www.un.org/sg/en/content/sg/statement/2021-08-09/secretary-generals-statement-the-ipcc-working-group-1-report-the-physical-science-basis-of-the-sixth-assessment?_gl=1%2Apg8kv3%2A_ga%2AMTQwODEzOTE5OS4xNzQyNDA0ODQz%2A_ga_TK9BQL5X7Z%2AMTc0MjQwNDg0My4xLjAuMTc0MjQwNDg0My4wLjAuMA..%2A_ga_S5EKZKSB78%2AMTc0MjQwNDg0My4xLjAuMTc0MjQwNDg0NS41OC4wLjA.
5. "UN Climate Report: It's 'Now or Never' to Limit Global Warming To 1.5 Degrees," United Nations, April 4, 2022, https://news.un.org/en/story/2022/04/1115452?_gl=1*7jypdu*_ga*MTQwODEzOTE5OS4xNzQyNDA0ODQz*_ga_TK9BQL5X7Z*MTc0MjQwNDg0My4xLjEuMTc0MjQwNTAyMy4wLjAuMA.
6. "Secretary-General Calls on States to Tackle Climate Change 'Time Bomb' through New Solidarity Pact, Acceleration Agenda, at Launch of Intergovernmental Panel Report," United Nations, March 2, 2023, https://press.un.org/en/2023/sgsm21730.doc.htm.

Notes

Chapter 1

1. Rachel Yehuda, "Transgenerational Effects of Posttraumatic Stress Disorder in Babies of Mothers Exposed to the World Trade Center Attacks during Pregnancy," *Journal of Clinical Endocrinology & Metabolism* 90, no. 7 (2005): 4115–118, https://doi.org/10.1210/jc.2005-0550.

2. Brene Brown, *Daring Greatly: How the Courage to Be Vulnerable Transforms the Way We Live, Love, Parent, and Lead* (Avery, 2015), 34.

3. Caroline Hickman et al., "Climate Anxiety in Children and Young People and Their Beliefs About Government Responses to Climate Change: A Global Survey," *The Lancet Planetary Health* 5, No. 12 (2021): e863–e873, https://doi.org/10.1016/S2542-5196(21)00278-3.

4. R. Eric Lewandowsk et al., "Climate Emotions, Thoughts, and Plans Among US Adolescents and Young Adults: A Cross-Sectional Descriptive Survey and Analysis by Political Party Identification and Self-Reported Exposure to Severe Weather Event," *Lancet Planetary Health* 8, no. 11, e879–e893, https://www.thelancet.com/journals/lanplh/article/PIIS2542-5196(24)00229-8/fulltext.

5. Yale Program on Climate Change Communication, "The Six Americas 2023: A National Public Opinion Survey on Climate Change," n.d., climatecommunication.yale.edu. Yale Program on Climate Change Communication, "Global Warming's Six Americas," December 14, 2023, https://climatecommunication.yale.edu/about/projects/global-warmings-six-americas/.

6. Lina Adil et al., "Climate Risk Index 2025," Germanwatch, February 12, 2025, https://www.germanwatch.org/sites/default/files/2025-02/Climate%20Risk%20Index%202025.pdf.

7. Yoko Nomura et al., "Prenatal Exposure to a Natural Disaster and Early Development of Psychiatric Disorders During the Preschool Years: Stress in Pregnancy Study," *Journal of Child Psychology and Psychiatry* 64, No. 7 (2023): 1080–091, doi.org/10.1111/jcpp.13698.

8. Vincent A. Landau et al., *Analysis of the Disparities in Nature Loss and Access to Nature. Technical Report,* Conservation Science Partners, May 29, 2020, https://www.csp-inc.org/public/CSP-CAP_Disparities_in_Nature_Loss_FINAL_Report_060120.pdf.

Chapter 2

1. Olivia Sappenfield et al., "Adolescent Mental and Behavioral Health, 2023," in *National Survey of Children's Health Data Briefs,* Health Resources and Services Administration, 2024, www.ncbi.nlm.nih.gov/books/NBK608531/.

Notes

2. Ariella Cook-Shonkoff, "Here's How to Talk to Kids about Climate Anxiety," *Grist*, February 23, 2021, https://grist.org/climate/parent-therapist-how-to-talk-about-climate-change-anxiety-with-kids/.

3. Daniel Siegel and David Rock, "The Healthy Mind Platter for Optimal Brain Matter," Mind Your Brain, Inc., June 11, 2011, https://drdansiegel.com/healthy-mind-platter/.

4. Xiaohe Ren et al., "A Systematic Review of Parental Burnout and Related Factors Among Parents," *BMC Public Health* 24, no. 1 (2024): 376. https://doi.org/10.1186/s12889-024-17829-y.

5 Lucy McBride, "By Now, Burnout Is a Given," *The Atlantic*, June 30, 2021, https://www.theatlantic.com/ideas/archive/2021/06/burnout-medical-condition-pandemic/619321/.

6. Gun Violence Archive, "GVA—10 Year Review," accessed March 24, 2025, https://www.gunviolencearchive.org.

Chapter 3

1. In the Making, "Living in the Metacrisis with Jonathan Rowson," YouTube Video, October 19, 2023, https://www.youtube.com/watch?v=IjOQB608ylQ.

2. Sally Weintrobe, *The Psychological Roots of the Climate Crisis* (Bloomsbury Academic, 2021).

3. Erich Fromm, *The Art of Loving* (Harper & Row, 1956).

4. Saundra Dalton-Smith, "The Real Reason You Are Tired and What to Do About It," TEDxAtlanta, March 2019, https://www.ted.com/talks/saundra_dalton_smith_the_real_reason_why_we_are_tired_and_what_to_do_about_it.

5. Yong Liu et al., "Prevalence of Healthy Sleep Duration among Adults—United States, 2014," *Morbidity and Mortality Weekly Report* 65, no. 6 (2016): 137–41, https://doi.org/10.15585/mmwr.mm6506a1.

6. A. G. Wheaton et al., "Short Sleep Duration Among Middle School and High School Students—United States, 2015," *Morbidity and Mortality Weekly Report* 67, no. 3 (2018): 85–90, http://dx.doi.org/10.15585/mmwr.mm6703a1.

7. U.S. Department of Housing and Urban Development, Office of Community Planning and Development, *The 2024 Annual Homelessness Assessment Report (AHAR) to Congress,* December 2024, https://www.huduser.gov/portal/sites/default/files/pdf/2024-AHAR-Part-1.pdf.

8. Pema Chödrön, *When Things Fall Apart: Heart Advice for Difficult Times* (Shambhala, 2000).

Notes

9. Shelley E. Taylor, "Tend and Befriend Biobehavioral Bases of Affiliation Under Stress," *Current Directions in Psychological Science* 15, no. 6 (December 2006), https://doi.org/10.1111/j.1467-8721.2006.00451.x.

Chapter 4

1. Organization for Economic Cooperation and Development (OECD), "PF2.1. Parental Leave Systems," *OECD Family Database*, February 2024, https://www.oecd.org/content/dam/oecd/en/data/datasets/family-database/pf2_1_parental_leave_systems.pdf.

2. Sally Weintrobe, *The Psychological Roots of the Climate Crisis* (Bloomsbury Academic, 2021).

3. Vincent J. Felitti et al., "Relationship of Childhood Abuse and Household Dysfunction to Many of the Leading Causes of Death in Adults. The Adverse Childhood Experiences (ACE) Study," *American Journal of Preventive Medicine* 14, no. 4 (1998): 245–58, https://doi.org/10.1016/s0749-3797(98)00017-8.

4. Felitti et al., "Relationship of Childhood Abuse."

5. Ariella Cook-Shonkoff, "Parenting in the Age of 'Eco-Anxiety': Wildlife Fears, and a Deeper Dread," *Washington Post*, December 5, 2019, https://www.washingtonpost.com/lifestyle/2019/12/05/parenting-age-eco-anxiety-wildfire-fears-deeper-dread/.

6. Oliver Milman, "'Magical Thinking': Hopes for Sustainable Jet Fuel Not Realistic, Report Finds," *The Guardian*, May 14, 2024, https://www.theguardian.com/environment/article/2024/may/14/sustainable-jet-fuel-report.

7. Anthony Leiserowitz et al., "Climate Change in the American Mind: Beliefs & Attitudes, December 2022," Yale Program on Climate Change Communication, February 16, 2023, https://climatecommunication.yale.edu/publications/climate-change-in-the-american-mind-beliefs-attitudes-december-2022/.

8. Morning Consult, "Parents and the Planet: How Sustainability Impacts Purchasing Decisions," *HP 2023 Sustainable Purchasing Pulse*, 2023, https://press.hp.com/content/dam/sites/garage-press/press/press-kits/2023/2022-sustainable-impact-report/HP%20Sustainable%20Purchasing%20Pulse%202023_Executive%20Summary.pdf.

Chapter 5

1. Joshua M. Schrei, host, *The Emerald*, podcast, season 1, episode 76, "The Revolution Will Not Be Psychologized, Part 2 (Interview w/Báyò Akómoláfé)," April 7, 2023, https://theemeraldpodcast.buzzsprout.com/317042

Notes

/episodes/12607531-the-revolution-will-not-be-psychologized-part-2-interview-w-bayo-akomolafe.

2. Jia Tolentino, "What to Do with Climate Emotions?" *New Yorker*, July 10, 2023, https://www.newyorker.com/news/annals-of-a-warming-planet/what-to-do-with-climate-emotions.

3. Jonathan Haidt, *The Anxious Generation* (Penguin Press, 2024), https://strategylab.ca/wp-content/uploads/2024/07/The-Anxious-Generation-Supplemental-Resources.pdf.

Chapter 6

1. Tara Brach, *Radical Acceptance: Embracing Your Life with the Heart of a Buddha* (Bantam, 2004).

2. Deutsches Aerzteblatt International, "The Healing Powers of Music: Mozart and Strauss for Treating Hypertension," *ScienceDaily*, June 20, 2016, https://www.sciencedaily.com/releases/2016/06/160620112512.htm.

3. Jonathan Haidt, *The Anxious Generation* (Penguin Press, 2024), https://strategylab.ca/wp-content/uploads/2024/07/The-Anxious-Generation-Supplemental-Resources.pdf.

4. Robin Wall Kimmerer, *Braiding Sweetgrass: Indigenous Wisdom, Scientific Knowledge, and the Teachings of Plants* (Milkweed Editions, 2015).

5. Sara H. Konrath et al., "Changes in Dispositional Empathy in American College Students Over Time: A Meta-Analysis," *Personality and Social Psycholgy Review* 15, no. 2 (2011):180–98, https://doi.org/10.1177/1088868310377395.

6. Fan Wang et al., "A Systematic Review and Meta-Analysis of 90 Cohort Studies of Social Isolation, Loneliness and Mortality," *Nature Human Behaviour* 7 (2023):1307–319, https://doi.org/10.1038/s41562-023-01617-6.

7. Julie Early Sifuentes, Emily York, and Courtney Fultineer, "Social Resilience and Climate Change: Findings from Community Listening Sessions," *The Lancet Planetary Health* 5, no. S3 (2021), doi:10.1016/S2542-5196(21)00087-5.

8. Julie Early Sifuentes et al., "Social Resilience and Climate Change: Findings from Community Listening Sessions," *The Lancet Planetary Health* 5, no. S3 (2021), https://doi.org/10.1016/S2542-5196(21)00087-5.

9. Neil E. Klepeis et al., "The National Human Activity Pattern Survey (NHAPS): A Resource for Assessing Exposure to Environmental Pollutants," *Journal of Exposure and Environmental Epidemiolgy* 11, no. 3 (2001): 2312–352, https://doi.org/10.1038/sj.jea.7500165.

Notes

10. Gregory N. Bratman et al., "Nature Experience Reduces Rumination and Subgenual Prefrontal Cortex Activation," *PNAS* 112, no. 28 (2015): 8567–572, https://doi.org/10.1073/pnas.1510459112.
11. Joshua Abraham Heschel, *Prayer and Radical Amazement* (Yale University Press, 2021).
12. Thomas Doherty et al., "Environmental Identity-Based Therapies for Climate Distress: Applying Cognitive Behavioural Approaches," *Cognitive Behaviour Therapist* 17 (2024): e29. doi:10.1017/S1754470X24000278.

Chapter 7

1. Jonathon Haidt, *The Righteous Mind: Why Good People Are Divided by Politics and Religion*, (Vintage, 2013).

Chapter 8

1. Tricia Taormina, "Climate Moms to the Rescue!" *Elle*, October 20, 2020, www.elle.com/culture/a34525892/climate-moms-to-the-rescue/.
2. Erik Erikson, *Childhood and Society* (W. W. Norton, 1993).
3. Susan Clayton and McKenna F. Parnes, "Anxiety and Activism in Response to Climate Change," *Current Opinions in Psychology* 62 (2025), https://doi.org/10.1016/j.copsyc.2025.101996.
4. Ro Randall, "Ro Randall on the Psychology of Climate Change Concern and Action," *Bridging the Carbon Gap*, podcast, April 21, 2022, https://podcasts.apple.com/ca/podcast/ro-randall-on-the-psychology-of-climate-change/id1630689195?i=1000567075178.

Chapter 9

1. Climate Action Families, "Our History," About Us, December 28, 2024, https://climateactionfamilies.org/about/.
2. Sarah Spengeman, host, "Talking with Lucy Molina and Shaina Oliver Molina Interview," season 2, episode three, *Listen Notes*, April 23, 2022, 48 mins., 42 sec., https://www.listennotes.com/podcasts/hot-mamas/season-2-episode-three-oNj0tzFvs4P/.
3. Environmental Defense Fund, "No More Sacrifice Zones: Stopping Pollution from Suncor Refinery | Shaina Oliver's Story," YouTube Video, May 4, 2022, https://www.youtube.com/watch?v=FxymFYkiwQ0.
4. Katherine Hayhoe, "The Most Important Thing You Can Do to Fight Climate Change: Talk About It," TEDWomen Video, November 2018, https://www

Notes

.ted.com/talks/katharine_hayhoe_the_most_important_thing_you_can_do_to_fight_climate_change_talk_about_it?language=en.

5. Roman Krznaric, "How to Be a Good Ancestor," TED Video, October 2020, https://www.ted.com/talks/roman_krznaric_how_to_be_a_good_ancestor?language=en.

Chapter 10

1. Benjamin Barber, "Jihad vs. McWorld," *The Atlantic,* March 1992, https://www.theatlantic.com/magazine/archive/1992/03/jihad-vs-mcworld/303882/.

2. Per Espen Stoknes, "The Coming Climate Disruptions: Are You Hopeful?" *Psychology Today,* April 6, 2015, https://www.psychologytoday.com/us/blog/what-we-think-about-global-warming/201504/the-coming-climate-disruptions-are-you-hopeful.

3. Rebecca Solnit, *Hope in the Dark* (Haymarket Books, 2016).

4. Christiana Figueres, "The Case for Stubborn Optimism on Climate," TED Video, October 2020, https://www.ted.com/talks/christiana_figueres_the_case_for_stubborn_optimism_on_climate.

5. Louise Chawla, "Childhood Nature Connection and Constructive Hope: Helping Young People Connect with Nature and Cope with Environmental Loss," in *High-Quality Outdoor Learning,* eds. Rolf Jucker and Jakob von Au, (Springer, Cham, 2022), 95–122, https://doi.org/10.1007/978-3-031-04108-2_5.

Index

abortion, 22
Abrams, David, 9
activism. *See* climate action and activism
adaptive parenting, 81–84
adaptivity, 149–157
addiction, 111–113, 116
Adverse Childhood Experiences (ACEs), 100–101
Akomolafe, Bayo, 128
Albrecht, Glen, 228
alloparents, 95, 190
Allured, Elizabeth, 5
ambiguous loss, 131
Amini, Fari, 53
ancestral wisdom, 191–192
anthropocene, 31, 228
anti-doomer resources, 320–321
Anti-Doomer Sweet Spot, 298–302
anxiety, 41–43
   Climate Anxiety Tonic, 56–57, 65
   co-regulation and, 52–54
   eco-anxiety, 3–4, 43, 131, 132
   Macro stressors, 41–42, 45–46, 63, 73, 94, 141
   managing parental anxiety, 54–56
   Micro stressors, 42, 46, 55
   parenthood and, 3
   politics and, 57–58
   tools for, 311–312
   trickle-down theory of, 43–52
   See also climate anxiety
artificial intelligence (AI), 6
art shaming, 158
art therapy, 10, 86, 160–161, 197, 199. *See also* Creative Exercises
attachment figures, 53
attitude, 32, 101, 157, 240, 278, 296
automatic negative thoughts (ANTs), 155–156, 307
autonomic nervous system, 127, 177–178
awe, 163, 166, 194–195, 198–199
Ayurveda, 188

babymoon, 19
baby showers, 18, 19
Barber, Benjamin, 292
Bastida, Xiye, 274–275
Batuyev, Maksim, 282
behavioral contagion, 107–108
Bezos, Jeff, 283

Index

biomimicry, 303
biophilia, 33, 131
BIPOC (Black, Indigenous, and People of Color) communities, 11–12, 33, 280, 317–318
Black Lives Matter, 266, 272
blessingway ceremony, 19
body intelligence, 172
body map. *See* Existential Body Map
Boomers, 262
boredom, 87, 154
Boyd, Andrew, 174
Brach, Tara, 135, 143, 151, 153, 154, 312
brain-gut axis, 177–178
Brave Bull Allard, LaDonna, 273
breathing exercises, 69–70, 180, 302
British Petroleum (BP), 96
Brown, Brene, 25–26, 158
Buddhism, 64–65, 82, 189, 207
   beginner's mind, 65
   interbeing, 189
   loving-kindness, 27
   Tibetan Buddhism, 161
   Zen Buddhism, 162, 172
bullying, 7
burnout, 55, 94, 147, 186, 247–248
Buzzell, Linda, 195

capitalism, 33, 94. *See also* consumption and consumerism
Carbon Conversations, 250
carbon footprint, 95, 96, 111
Chellun, Chryso, 275
Chesapeake Climate Action Network, 270–271
Chibanda, Dixon, 280

child abuse and neglect, 45, 100
childbirth, 19, 20, 21, 23, 189
Chödrön, Pema, 82
climate action and activism, 44–45, 103, 165, 247, 252–253. *See also* intergenerational climate action
Climate Action Families (Seattle group), 260
Climate Action Venn Diagram, 248–250
climate anxiety, 42–50, 108, 109
   compartmentalizing, 12
   eco-anxiety compared with, 43, 131
   privilege and, 12
Climate Anxiety Cocktail, 46, 59, 295
Climate Anxiety Tonic, 56–57, 65
Climate Aware Therapist Directory, 108, 322
climate-aware therapists, 5, 42, 133, 166
climate change
   anthropocene, 31, 228
   corporate greenwashing, 102–103
   IPCC reports, 8, 11, 46, 95
   overshoot, 31
   statistics, 7–8
   tipping points, 8, 31
Climate Change Theatre Action, 275–276
climate distress, 42, 43, 126, 131, 133, 195
climate dread, 43, 131
Climate Emotion-Action DNA Model, 245–248
climate emotions, 131–134
   climate activism compared with, 247

Index

climate dates and, 117
doomerism versus, 295
intergenerational climate connectors and, 281
naming, 154
See also anxiety
Climate Emotions Layered Mandala, 139–140
Climate Emotions Pellet, 134–136
Climate Emotions Wheel, 137–139, 154, 241
climate empathy, 131, 228
climate intergenerational spaces, 321
Climate Manifesto, 305–306
Climate Mental Health Network, 137, 266, 282
Climate Museum, 276
climate parent groups, 321
climate psychology resources, 322
Climate Recipes for Relief, 59–60
Closing the Bones ceremony, 19
cognitive dissonance, 3–4, 103–105, 110, 132, 141, 171–172
colonialism, 33, 65, 169
communication, 206–207
  appropriate content, 212–213
  determining readiness, 208
  difficult conversations, 232–235
  framing information, 227–232
  How to Talk to Kids About Climate Change Cheat Sheet, 209–211
  listening, 233–234
  nonviolent communication (NVC), 234
  timing, 213–215
  voice, validate, and vision, 215–226

community-supported agricultural (CSA), 78, 79, 187
compartmentalization, 4, 32, 47, 54
connection, 188–193
consumption and consumerism, 12, 64, 102, 103, 151, 303
co-regulation, 52–54
cortisol, 20, 77
COVID-19 pandemic, 2
  adaptation and adaptive parenting, 63, 83
  family dance parties, 163
  hugging and connection, 191
  long-term COVID-19, 131
  online yoga classes, 182
  Our Words Collide (documentary), 166
  parental burnout, 55
  post-pandemic digital acceptance, 143–145
Creative Exercises, 10
  Anti-Doomer Photo Project, 235–236
  Climate Recipes for Relief, 59–60
  Digging Into Your Climate Disconnect, 120–121
  Eco-Ancestral Identity Timeline, 201–203
  Home Sweet Home, 88–89
  Stained Glass Emotions, 145–146
  Uniting Hearts and Hands Group Mural, 289–290
  Visualizing the Future with a Solution-Focused Mandala, 255–256
  Who Am I? Parent Collage, 38–39

Index

creativity, 86–88
  art shaming, 158
  in Existential Body Map, 157–167
  mandalas, 161–162
  in music and dance, 162–165
  new creative frontiers, 162–165
  reclaiming, 158–160
  in stories, acting, and writing, 165–167
cultural appreciation, 9, 186
cultural appropriation, 9, 148, 169, 186

Dalton-Smith, Saundra, 66, 67
Davenport, Leslie, 47, 136
DearTomorrow, 93, 321
democracy, 6, 64, 261
denial, 97–98, 106
despair, 105–106
Dialectical Behavioral Therapy (DBT), 82
Diaz, Junot, 288
disavowal, 99
disenfranchised grief, 131–132
disinformation, 6
Doherty, Thomas, 201
domestic violence, 22, 76, 100, 101
Don't Look Up (film), 95, 320
doomerism, 295–301
dopamine, 53, 66

Earth Day, 160, 193, 286–287
eco-anxiety, 3–4, 43, 131, 132. *See also* climate anxiety
eco-art therapy, 197, 199. *See also* art therapy
ecofeminism, 33

eco-guilt, 132
Eco-Values Checklist, 79–81
Eisley, Loren, 304
embeddedness, 193–200
embodied emotions, 127
emotional-behavioral disconnect, 108–111
emotional regulation, 52, 180–181
empathy, 189, 221, 232–234
environmental racism, 12, 33, 262, 272. *See also* racism
Erikson, Erik, 17, 242
evolution, 55, 96–97, 99, 148, 190
evolutionary psychology, 96–97
Existential Body Map (EBM), 148–149, 199
  Adaptive (head), 149–157
  Connected (hands), 188–193
  Creative (heart), 157–167
  Embedded (human-environment interface), 193–200
  Grounded (feet), 176–188
  Intuitive (gut), 168–176

Family Eco-Values Checklist, 79–81
Family Pauses. *See* Parent and Family Pauses
feminism, 33, 58
fight, flight, or freeze, 52, 97, 296
Figueres, Christiana, 300
forest bathing (*shinrin-yoku*), 77, 196
fossil fuel industry, 96
Four Directions, 185–186
Frankenstein (Shelley), 31
Friedman, Russell, 142
Friendship Bench, 280–281
Fromm, Erich, 65–66

Index

Future Design movement, 266, 282, 284
Fuyani, Xoli, 285

Gaming for Justice, 275
generational divides, 260–264. *See also* intergenerational climate action
Generations United, 265
generativity, 17
Gen Z, 261, 266, 282, 298
Ginsburg, Kenneth, 73, 74
Global Climate Risk Index, 32
Global South, 32, 63
Godh Bharai, 19
Gonzales, Jose, 297
Goodall, Jane, 285
gratitude, 157, 182, 187, 195, 288, 313
Green New Deal, 273–274
greenwashing, 102–103, 300
grief, 20–24, 141–143
 birth-related grief, 21–22
 children's grief, 221
 climate change and, 108, 126, 165–166
 death and, 200
 disenfranchised grief, 131–132
 parenthood and, 20–22, 25
 remaining present during, 136
groundedness, 176–188
Gualinga, Helena, 274–275
gun violence, 58, 64, 147, 208, 215
 mass shootings, 7, 58, 104, 131, 141, 297
 school shootings, 16–17, 33, 104, 131
Gun Violence Archive, 58

Haidt, Jonathan, 144, 174, 232
Halifax, Joan, 301
Hayhoe, Katherine, 277, 320
Healthy Mind Platter, 49–50, 51
Held vs. Montana, 269
helicopter parenting, 17, 94–95, 229
HelpAge International, 265
Heschel, Abraham Joshua, 198–199
Hinduism, 19
shinrin-yoku (forest bathing), 77, 196
homeland, 62–63
hope, 293–295
hugging, 72, 77, 191
hyperobject, 95
hypocrisy, embracing, 119–120

identity, 237–238
 child self and, 239–242
 e-piphanies and aha! moments, 250–254
 Parent Identity Spiral, 34–37
 revisiting puberty, 242–244
 sculpting, 245–256
Indigenous-focused resources, 313–315
Indigenous wisdom, 184–186. *See also* Traditional Ecological Knowledge
inequity, 11, 34, 64, 94, 195
infertility, 22

Index

intergenerational climate action, 265–266, 281–287
  climate intergenerational spaces, 321
  courts and the law, 269–270
  creativity and, 275–276
  difficult conversations and, 277–281
  generational divides and, 260–264
  models of, 266
  parent-led advocacy, 270–274
  school-based intergenerational collaboration, 267–269
Intergovernmental Panel on Climate Change (IPCC), 8, 11, 46, 95
intuition, 168–176

James, John, 142
Jewish traditions and holidays
  Mikveh, 19
  Passover, 56–57
  shalom bayit ("peace of the home"), 73–74
  Sukkot, 188
Johnson, Ayana Elizabeth, 248
Johnstone, Chris, 294, 319
Juliana vs. U.S. Government, 269
Jung, Carl, 94, 161

Kamenetz, Anya, 141
Keshet, Mor, 141
Keys, Alicia, 297
Kimmerer, Robin Wall, 185, 314
Krznaric, Roman, 283–284, 286, 319
Kubit, Jill, 93

Lannon, Richard, 53
Las Madres del La Plaza de Mayo, 273
laughter, 174
Levine, Peter, 177
Lewis, Thomas, 53
Lifton, Robert J., 100
Lighthouse Parenting, 82
limbic resonance, 53
Lorax, The (Seuss), 264–265, 320
Lorde, Audre, 291
Louv, Richard, 33

Macro stressors, 41–42, 45–46, 63, 73, 94, 141
Macy, Joanna, 65, 207, 253, 287, 294, 319
magical thinking, 106–107, 108
Mailer, Becky, 275
mandalas, 161–162
  Climate Emotions Layered Mandala, 139–140
  Visualizing the Future with a Solution-Focused Mandala, 255–256
March for Our Lives, 272
March on Washington, 267
Markey, Ed, 274
mass shootings, 7, 58, 104, 131, 141, 297. *See also* gun violence
Maté, Gabor, 23, 101
Mayfield, John, 172
McBride, Lucy, 55
McGrath, Carolyn, 268–269
McKay, Adam, 95
metacrisis, 63–64
MHW, defined, 9
micromanaging, 41, 95

Index

Micro stressors, 42, 46, 55
Milagro, Lil, 275, 283
Milne, A. A., 105
miscarriage, 20, 22, 23, 141
Miyawaki, Akira, 285
modern autopilot (MA), 151–153, 187, 251, 265, 293
Molina, Luz E. "Lucy," 271–272
Molina-Moton, Elijah, 272
Montgomery Bus Boycott, 267
moral injury, 62, 132
Morton, Timothy, 95
Moser, Susan, 247
Mothers Rise Up, 275
motivational interviewing (MI), 116–118
Musk, Elon, 296–297

nature appreciation, 195, 290
nature-deficit disorder, 33
Navajo traditions, 19
negativity bias, 85, 156, 227
neoliberalism, 64
neuroscience, 49–50, 52, 176
Nicholas, Kimberly, 8
9/11, 20, 166
NorCal Climate Elders, 281–282
nurture nature family philosophy, 77–81

Ocasio-Cortez, Alexandria, 274
Odell, Jenny, 183
Olson, Julia, 269
optimism, 294, 300
other-care, 192–193
Our Kids' Climate, 93, 260, 321

parental burnout, 55, 94, 147, 186, 247–248
Parent and Family Pauses, 10–11
  Climate Action Venn Diagram, 249–250
  Climate Emotions Layered Mandala, 139–140
  Climate Emotions Pellet, 135
  Cognitive Dissonance, 104–105
  Communication, 217
  COVID-19 Pandemic, 83
  Drishti, 184
  Emotional-Behavioral Disconnect, 110–111
  Empathy, 233
  Family Eco-Values Checklist, 80–81
  Grief Graph, 142–143
  Heart-to-Heart Conversation With Your Youth Self, 239–241
  Land Sea Air Scan, 129–130
  modern autopilot (MA), 152
  Music and Creativity, 165
  Parent-Borne Confidence, 244
  Peace of the Home, 75
  Play, 176
  Self-Care Acronym, 199–200
  Self-Care Exercises, 51
  Shape-Shifting the Vulnerability-Responsibility Dialectic, 29–30
  Solution-Focused Thinking, 302
  Vulnerability, 26
  Wise Guides and Resilience, 170–171
Parent Centering Practice (PCP), 10, 11, 67–73, 82, 128, 140
Parent Identity Spiral, 34–37

Index

Parents For Future, 260, 321
Patil, Neelam, 285
patriarchy, 33, 58, 141
"peace of the home," 73–77
Pelosi, Nancy, 274
perfectionism, 84–85
perinatal period, 20–21
Perry, Bruce, 53
Pihkala, Panu, 137
play, child-centered, 174–175
polycrisis, 11, 41, 63–64, 93, 189, 239
polyvagal theory and exercises, 127, 151, 177–178, 311
Porges, Stephen, 177–178
postpartum ceremony, 19
postpartum health and recovery, 19–21
Postpartum Support International (PSI), 21
post-traumatic growth, 24
post-traumatic stress disorder (PTSD), 32, 34, 132, 196
Powell, Emma, 275
pregnancy, 1, 3, 9, 16, 20, 21, 22, 23, 32
preppers, 296–297
pre-term births, 22, 32
pre-traumatic stress, 131–132
psychic numbing, 99–100
puberty, 242–244

racism, 57, 64, 94, 147, 215
   environmental racism, 12, 33, 262, 272
   standing up to, 18
   structural racism, 96, 100
   workplace, 131

RAIN meditation, 154–155, 157, 312
Randall, Rosemary, 250–251
Randolph, A. Philip, 266
rationalization, 102–103
Research Institute for Future Design, 266
resilience, 304–305
rest and sleep, 65–67
Ridwell, 107
Rock, David, 49–50
Rosenberg, Marshall, 234
Rowson, Jonathan, 64

school shootings, 16–17, 33, 104, 131. *See also* gun violence
screens, 66, 87, 143–144, 175–179
seasonal transitions, 186–188
self-care, 50–51, 53, 67–69, 140, 181, 247. *See also* Parent Centering Practice
Self-Care Acronym, 199–200
Self-Care Weekly Tracker, 72, 309
self-harm, 32, 48, 126, 144
Serenity Prayer, 136
serotonin, 77, 197
Seuss, Dr., 264–265, 320
Seventh Generation Principle, 185, 266, 313
sexual abuse, 22
shadow theory, 94, 115
shame, 25–26
   birth-related trauma and, 23
   change and, 244
   communication and, 207, 215, 216, 230, 231
   criticism and, 85
   grief and, 141

perfectionism and, 84
secrecy and, 207, 230, 231
vulnerability and, 25–26
Shelley, Mary, 31
Shugarman, Harriet, 227
Siegel, Daniel, 49–50, 154
silent dance parties, 259, 288–289
sleep and rest, 65–67
Slow Sundays, 74
snowplow parenting, 94–95
Socrates, 16
Sofer, Oren Jay, 234, 320
soft denial, 98
solastalgia, 132
soliphilia, 132
Solnit, Rebecca, 294
Somatic Experiencing, 127, 151, 177
Somatic Sensations Words, 138–139, 241
Stages of Change model, 111–113, 119
Standing Rock, North Dakota, 266, 273
stillbirth, 20, 22. *See also* miscarriage
Stokes, Mimi, 166
storytelling, 287
stress response system, 52, 97, 132, 247, 296
substance abuse, 22, 34, 48, 101
suicide and suicidality, 32, 34, 58, 98, 101, 126, 133, 144
Sunrise Movement, 274
Superstorm Sandy, 32
Suring, Ella, 267–268
symbiocene, 228

Taylor, Shelley E., 83
Thunberg, Greta, 252–253, 261, 282, 291, 319, 320

tiger parenting, 94–95
Tolentino, Jia, 136
toxic stress, 132
Traditional Ecological Knowledge (TEK), 313–314
trauma, 20–24
Adverse Childhood Experiences (ACEs) and, 100–101
ancestral wisdom and, 191
art therapy for, 158
attachment figure and, 53
birth-related trauma, 20–23
child-adult integration and, 238, 239
COVID-19 pandemic and, 143
creative trauma, 86
development trauma, 32
environmental and climate trauma, 32, 45, 49
intuition and, 168–169
memories and, 27
mind-body modalities and, 127
9/11 and, 166
parenthood and, 20–24
"peace of the home" and, 76–77
post-traumatic growth, 24
post-traumatic stress disorder (PTSD), 32, 34, 132, 196
pre-traumatic stress, 131–132
professional help for, 145
psychic numbing and, 99–100
Somatic Experiencing for, 177
Tools for Managing Anxious or Traumatized Nervous Systems, 311–312
trauma-informed yoga, 182
types of, 100–101
vicarious trauma, 132

trees, 193, 196–197, 303
Trump, Donald, 6, 57–58, 274, 285
Tsui, Tori, 294

vagus nerve, 127, 177, 302
volunteering, 192–193
vulnerability, 24–34

Ward, Maya, 275
Weintrobe, Sally, 64, 99, 303
wildfires, 3–4, 48–49, 99, 224–226
  Camp Fire (2018), 4
  smoke from, 3–4, 32, 104, 251–252, 292
Wilson, E. O., 33

Wilson, Kimberly, 178
Wirth, Kelsey, 237
wise guide, 169–171
Wizard of Oz, The (film), 113–115. *See also* Yellow Brick Road of Climate Behavioral Change

Yellow Brick Road of Climate Behavioral Change, 111–115, 120, 121
yoga, 70, 174, 181–183, 197, 199, 200

Zones of Regulation, 179–180, 312
Zuckerberg, Mark, 297